Treetop Books

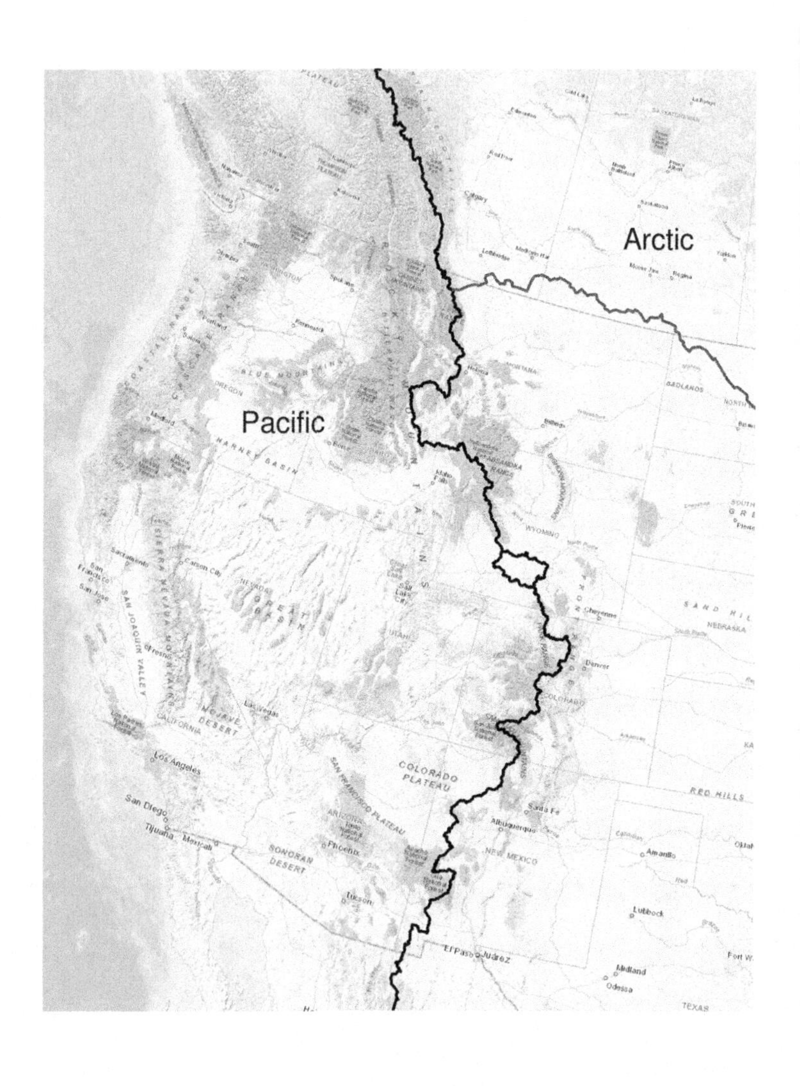

Pacific
Arctic
Edmonton
Red Deer
Calgary
Lethbridge
Medicine Hat
Saskatoon
Moose Jaw
Regina
Spokane
Portland
Salem
BLUE MOUNTAINS
OREGON
Medford
HARNEY BASIN
Helena
MONTANA
Billings
BADLANDS
WYOMING
Sacramento
San Francisco
San Jose
Carson City
NEVADA
GREAT BASIN
SIERRA NEVADA MOUNTAINS
SAN JOAQUIN VALLEY
Fresno
Salt Lake City
UTAH
Cheyenne
SAND HILLS
NEBRASKA
Denver
COLORADO
Las Vegas
MOJAVE DESERT
CALIFORNIA
Los Angeles
San Diego
Tijuana
Mexicali
COLORADO PLATEAU
SAN FRANCISCO PLATEAU
ARIZONA
Phoenix
SONORAN DESERT
Tucson
Santa Fe
Albuquerque
NEW MEXICO
RED HILLS
Amarillo
Lubbock
El Paso-Juárez
Midland
Odessa
TEXAS

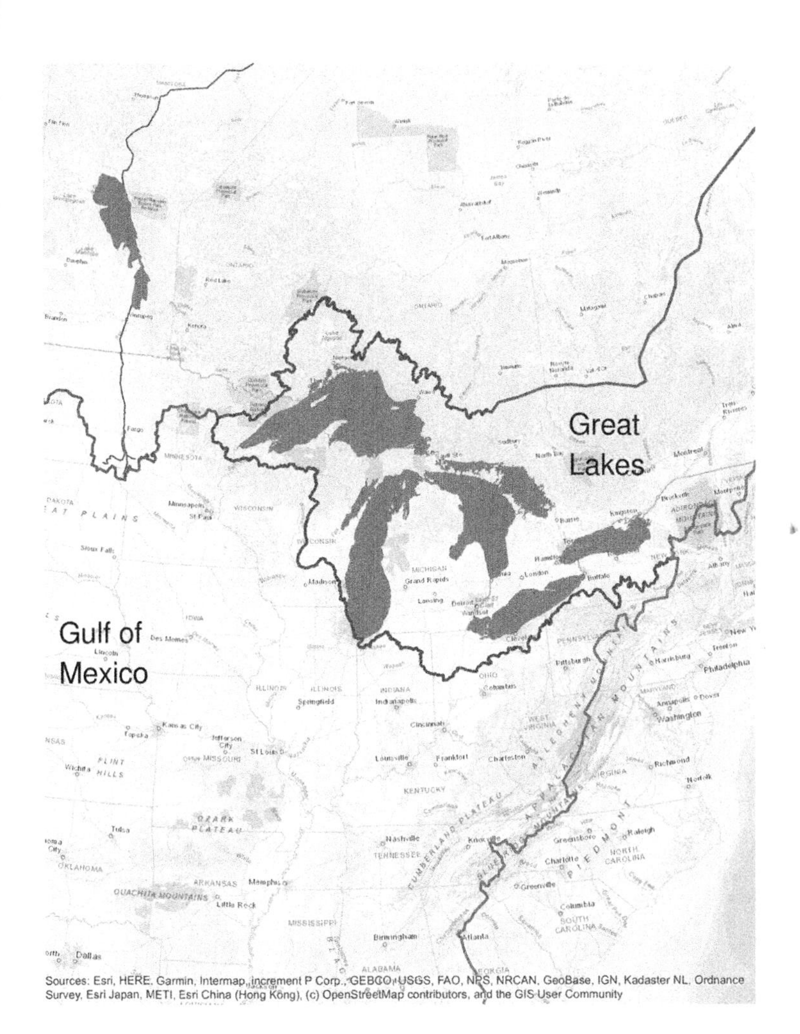
Great Lakes
Gulf of Mexico
Sources: Esri, HERE, Garmin, Intermap, increment P Corp., GEBCO, USGS, FAO, NPS, NRCAN, GeoBase, IGN, Kadaster NL, Ordnance Survey, Esri Japan, METI, Esri China (Hong Kong), (c) OpenStreetMap contributors, and the GIS User Community

LEAVING THE SUBURBS

A cross-continental journey through the front lines of the climate crisis

John E. Duke

Published by Treetop Books, Atlanta, Georgia
www.treetop-books.com

Second edition

ISBN-13: 978-0-9600089-4-0 (paperback)
ISBN-13: 978-0-9600089-5-7 (ebook)

Edited and designed by Backspace Ink
www.backspaceink.com

Except for celebrities, the names of the people in these stories have either been changed entirely or first names only may have been used on some occasions.

All maps and photographs created by the author.

Artwork credits:
Chapter 1: Marianne Lecesne / Anne McCallum
Chapter 2: author (map)
Chapter 3 / 4 / 5: Anne McCallum
Chapter 6 / 7 / 8: author
Chapter 9: Enphase Energy / PlugShare

Contents

List of Acronyms and Abbreviations

A/C	air conditioning
ANWR	Arctic National Wildlife Refuge
BMR	basal metabolic rate
DOT	Department of Transportation
EV	electric vehicle
FEMA	Federal Emergency Management Agency
HJAIA	Hartsfield-Jackson Atlanta International Airport
ICE	internal combustion engine
kW	kilowatt
kWh	kilowatt hour
MARTA	Metropolitan Atlanta Rapid Transit Authority
MEAG	Municipal Electric Authority of Georgia
MPG	miles per gallon
MW	megawatt
UP	Upper Peninsula
ppm	parts per million
USGS	United States Geological Survey

Introduction

THIS STORY BRIEFLY TOGGLES between a past event and the near-present moment—the past event being a climate history of the landscape on which the present event occurs: a solo bicycle ride through four major continental watersheds, 11 states, and four counties with so few people—less than two per square mile—that they would be considered "frontier," according to standards established by the US Census Bureau. I rode nearly 3,300 miles, and there were many days I traveled without ever seeing a traffic light. I saw the landscape slowly change from coastal plain to mountain to desert to vast horizons and back to forest again.

It took me two weeks to navigate across one continental watershed that had been entirely covered by a mile-thick ice sheet a mere 20,000 years ago. One massive glacier had lain right there and, except for the scattered pothole lakes used for recreation, everyone seemed too busy to notice the telltale signs: the earthen scars, the piles of rock left behind, and the erratic boulder here and there. It then took 7,000 years for Earth's temperature to slowly rise 9 degrees while that ice sheet retreated to the North Pole. And it took us less than a generation to raise it another 2 degrees (and it is still rising) from the burning of fossil fuels—lightning speed compared to geological time.

The near-present moments are also a journalistic reflection on that phenomenon (see "References and Notes" on page 167). It seems strange to me that we would let Earth burn, and I have spent the better part of my lifetime fighting for the root solutions to climate justice. This is my story on that journey.

Much has changed on the front lines of the modern-day climate saga since the first edition of this book was published three years ago. I bought an electric vehicle (EV) when the Chevy Bolt came out in 2016, and it soon became *Motor Trend*'s 2017 Car of the Year. It was a game changer. Before that, my only tool was to implore people to walk or catch the bus. I waited a while though and purchased a slightly used one in order to take advantage of the discounts in the robust Bolt used-car market that had already been established by 2019.

And the electric co-op where I live—the Municipal Electric Authority of Georgia (MEAG)—now beats the national average two times over in emissions-free power production, thanks in large part to an outsized investment in nuclear power. In fact, two-thirds of the *new* nuclear power that MEAG owns (in a 30-year-old nuclear power plant that is expanding) will be sold to the City of Jacksonville, Florida—for at least a decade—to help them lower their emissions. Meanwhile, the national average is now about 40 percent emissions-free with about half of that coming from nuclear.

One of my neighbors is a regional manager for MEAG, and he says that solar power is now a cheaper source of electricity than even natural-gas power plants. In fact, MEAG had recently signed a contract with a company called Pineview Solar to provide 80 megawatts (MWs) of solar power during peak times—that is 80 MWs at any one moment in time when demand is high, typically during the day and especially in the summertime. That would power about 8,000 homes when the sun is shining. The way battery prices are falling, that power could also be stored in utility scale batteries for steady release. Unfortunately, MEAG has no plans yet for a "big" battery, but my neighbor said the on-demand solar power was a huge bargain at less than 3 cents a kilowatt hour (kWh).

"We thought it was a typo," he said.

Some electric providers, and especially private companies, however, *are* investing heavily in battery plants because the price of batteries has fallen so dramatically that it is now cheaper to overbuild solar and store the excess than to continue running the old, high-maintenance fossil infrastructure. For 2021, 80 percent of all power-generating capacity added to the US grid will be from solar, wind, and/or accompanying battery facilities, according to the US Energy Information Administration.[1]

Since the first edition of this book, I took matters in my own hands, so to speak. I cajoled three other companies to join mine to put in curbs, demolish a bit of asphalt to make room for some topsoil, and landscape a sidewalk project that the city had left undone for years on a major corridor in my community with restaurants and coffee shops, leading to the Delta Air Lines headquarters. (We are nudging them to at least decarbonize their ground operations.)

Another citizen also took matters into his own hands and put a permanent bench at a popular bus stop on Main Street. The concrete pad was already there, so a group of us bought a shelter to go along with the bench and bolted it down one Sunday morning. Random acts of cheap, sustainable, low-maintenance, infrastructure. We communicated with nearby business owners but otherwise thought it would be easier to ask for forgiveness than it would be to ask for permission.

What hasn't changed since the last edition is the personal responsibility I feel that we are now in charge to handle the climate crisis. There is still a long way to go, and I won't be so dramatic as to say the challenge threatens our very existence, but it at least feels like it might threaten our existence *without chaos*. So, while I still carry this anxiety with me like a heavy piece of luggage (or should I say a loaded-down bicycle), I hope you enjoy the journey. Thank you for reading.

—John E. Duke

1 Glacier National Park

Take the long way home.

I KNOW WHAT IT is like to be a rock star without actually being a rock star. There were complete strangers asking me to pose with them for the camera. It was usually the person's spouse or mom or brother or whoever that snapped the picture with a Canon AE-1. I would smile and lean with one leg crossed and my arm resting on their shoulder, and for a brief moment it was as though we were best friends.

I had just pedaled halfway up the western slope of the continental divide in Glacier National Park on a bicycle. The traffic on the Going-to-the-Sun Road (Sun Road) in the summertime was so congested that a curfew (from 10 a.m. to 4 p.m.) had been imposed on bicycles. Therefore, I was pulled over in The Loop, a parking lot built inside the pocket formed by a hairpin curve, right at about 9:59. I think people had seen bicyclers before but not ones riding a bike that was loaded down with 30 pounds of gear, and not on a 17-foot-wide strip of asphalt, hanging on a cliff jammed against a vertical granite wall.

Whatever it was, the motorists jumped out of their cars in fascination at my endeavor. It was the third day of July, and dripping ice still hung on the north-facing sides of the wall—a wall so close you could reach out and touch—as the road cut in and out on a switchback on its way to the pass. The sweet aroma of conifer trees filled the subtle, breezy mountain air under a clear, blue sky only visible in long, thin slivers between the peaks. The ice on the wall seemed oddly out of place as the warm sun gently beat down on me, and a tiny bead of sweat popped out on my forehead.

I swear I could feel cool air release from the wall as the frozen water absorbed the day's energy and turned to a flowing liquid. This little part of the continent seemed to be declaring that this was the day—finally *the* day, *the* moment—that spring transitioned into summer this year. The other side of the road from the wall was thin air: a 1,000-foot cliff. And it was as though you were a slow-moving pin ball, bumping through a narrow corridor and hemmed in between rock and guard railings.

I was on summer break from college, on my way home after visiting relatives who lived on the west coast. Since leaving Everett, Washington, I was crossing my second mountain range, entering a second time zone, and had traveled nearly 600 miles. An automobile could have easily completed that distance in a day. I had already pedaled for 56 hours and had burned 36,400 calories. My friends thought that I would never make it back in time for fall classes because I would be stopping every few feet to pick up aluminum cans. Before I left, they had drawn on the chalkboard in the hallway of the dorm a stick image sitting on a bicycle, smiling, and tossing a can over his shoulder into a trailer being towed.

I had been harassing them to stop throwing away so much stuff and had started the first recycling program in the dorm. Today, recycling has become so mainstream that overwhelming peer pressure compels everyone to participate in communities all across the country. But are people buying all those recycled products? Some are completing the loop, and some aren't. The truth is that we can't recycle our way out of the biggest environmental crisis facing humanity, and so it seems recycling has now almost become a means to absolve us.[2]

At the time of my ride, though, I just thought, "How silly can they be? I don't have time for that."

Besides, I didn't have a trailer and I didn't have a support vehicle. I was riding alone and all my possessions were on a rack over the back wheel or in panniers hanging off of it. I had a piece of rope with which to form a centerline for the rain fly. I had a sleeping bag, a backpacking stove, a bottle of liquid fuel that lasted about eight days, and some food. I had a toothbrush, a bowl, a knife, a spoon, a change of socks and underwear, an extra T-shirt, and a fleece packed away for the cool nights. I wore spandex biking shorts, a T-shirt, a bike helmet, a side mirror attached to my eyeglasses, and flip-down sunglasses. Two water bottles were in hangers attached to the bike frame.

But my friends did know me well and called me "Nature Boy." I was in veterinary school at Auburn University, and I had contemplated dropping out many times. A confidante outside of my circle of academics convinced me that I could make an impact in the "big picture" through a public health career, so I stuck it on through at least for the time being.

During the last week of school during final exams, I read a treatise on the implications of the industrial revolution on the carbon cycle. I also had a knack for historical fiction, and another time during finals I had read *Gone with the Wind* by Margaret Mitchell for the second time in my life.

People stopped by my dorm room and asked incredulously, "Aren't you studying?"

I picked up *National Geographic*, *Outdoor Magazine*, and *Popular Science* whenever I had a few extra dollars. I read the newspaper, looking for the latest science and environmental reporting like I was digging for gold.

I made my classmates carpool wherever we couldn't walk. The intramural basketball coach asked me why I was so concerned about pollution when cars now had catalytic converters.

"It's not CO … it's CO *two*," I answered, long before Al Gore made his crusade. When people today tell me they don't like a politician—especially a Democrat who is telling them what to do—I say, "It was *me* who told *Al Gore*."

For some reason, they don't believe me. It could be that I have never met the man and I certainly didn't know who he was when I was explaining to my classmate exactly what catalytic converters do and don't do.

Caroline, my girlfriend and who was also in my proposed graduating class at Auburn, flew with me to Washington. As we sat in the Seattle–Tacoma International Airport, waiting for her return flight to board, she asked, "Are you sure you want to do this?"

"Do what?"

"Ride a bicycle home."

"Come on, Caroline. You know I have been planning and training for months."

"Okay. Promise you will call home often?"

"Promise."

We had a rocky relationship. She loved me, but I think she wanted someone more committed to a real career. Although I never confided in her, I knew that she knew I had other notions than the path I was on. You were supposed to be thankful for the opportunity to have a chance at a lucrative career like medicine or law. And I suppose that I was being irreverent and disrespectful. I didn't want to hurt her feelings at a time like this, so I just smiled and gently held her hand as I kept the thought to myself.

"Look, Caroline, it is my choice if I do not wish to finish school."

I estimated that I had a few days to think about it …

CAROLINE GREW UP IN one of the most well-to-do neighborhoods of Birmingham, Alabama. She grew up in one of those places where the real estate gets so high because of its close proximity to downtown yet somehow stayed away from the decay that crept into most urban areas of America during the latter half of the last century. Her neighborhood looked like everyone who lived there had their lawn manicured by servants. The school system was so good that everyone was expected to become a rocket scientist or a senator, if not a doctor or a lawyer. I, on the other hand, grew up in the western suburbs of Birmingham, where most people lived on a little more modest means and usually cut their own grass. The public education system was sufficient. We aspired to be athletes mostly, not scholars.

My childhood occurred during the time of rapid integration. I was born 804 days after Lyndon Johnson signed the Voting Rights Act of 1965. In 1972, my family moved into a shiny, brand new subdivision off of US Highway 78 where they were still clearing trees and framing houses. By the time I reached fifth grade, our neighborhood was nearly an even mix of Black and white families with a few South Vietnamese immigrants seeking asylum from the Vietnam War. I grew up in a high school where I was one of two white kids on the basketball team.

In the generation before me, my mom had graduated from the same school system in all white classes. Integration was still something of a novelty when we moved to our first home and, although my parents were kind and gracious people to everyone they met, they had no African American friends that I could remember. It might have baffled them—or at least been a curiosity—as to why I played basketball on street courts with all Black kids.

But there I was, and some of my friends sometimes yelled out, "Hey, Oreo!"

My mom was a devout Catholic who married a Baptist sharecropper's son from the rural cotton fields of north Alabama. My dad had had enough of farming and, as soon as he graduated from high school, moved to the city where he met my mom. Whereas Caroline lived among the upper-middle class due to her dad's education and high-profile career, my parents never went to college. By the time I was in the ninth grade, I found that I could get no help from them on my homework. Nonetheless, they were just as wise as anyone's parents—rich, poor, or somewhere in between. They were married for 53 years before my dad passed away.

By the grace of God, I was born to parents who stayed together, who cared so deeply that I become a well-educated citizen, and who had the means to provide me every opportunity to become that citizen. Dad went from a hired hand in a warehouse, to a buyer for the warehouse, to a salesman for many warehouses, and gradually clawed his way into the middle class. They knew their roots and, by God, I was never to buy anything that I couldn't pay for. They lived within their means and never paid a dime of interest on a credit card.

I never understood why the civil rights movement centered on my home state. Caroline never seemed to wonder about these kind of questions nor did any of my friends. But it was one of those curiosities that hung with me for a long time, like a sauce simmering on the back burner. After all, there was South Carolina, the first state to leave the Union after the election of Abraham Lincoln; there was Georgia, the last to be readmitted to the Union because they were so slow to ratify the 14th Amendment. Finally, many years later, the movie *Selma* came out, and that long-standing curiosity bubbled up again. I was there the first week it opened and, like in most of my basketball days, I was the only white dude around.

The Voting Rights Act of 1965 led to far more reaching consequences than just power at the ballot box. I am reminded of the scene in *To Kill a Mockingbird* when Atticus Finch is standing in a courtroom, trying to defend an innocent Black man in front of an all-white jury during the Jim Crow era. Although the ability to vote is powerful, the Voting Rights Act of 1965 did much more than guarantee suffrage to all Americans equally. It also fulfilled the promise that "the accused shall enjoy the right to … an impartial jury," as granted by the 6th Amendment to the US Constitution, because being registered to vote also put you on the list to be called for jury duty.

While watching *Selma*, another lightbulb went off: I was amazed at how strategic the Southern Christian Leadership Conference was in choosing venues for their civil rights campaign. They were very smart but also incredibly brave and courageous in their analysis of where to go. They chose places where there was controversy and contrast. The belligerence of Dallas County Sheriff Jim Clark towards anyone of color—an African American woman named Annie Lee Cooper in this specific instance—who dared to think they could register to vote within his jurisdiction did the trick. Martin Luther King, Jr., wasn't trying to change the hearts and minds of the white supremacist; he was trying to tug at the moral consciousness, and thus gain the support of, the white moderate.

The civil rights leaders were also righteous but not self-righteous. And they were forever hopeful. When Martin Luther King, Jr., said on the steps of the state capital at the end of the second attempt to march to Montgomery, "The arc of the moral universe is long but it bends toward justice," he concisely acknowledged a profound characteristic of the faith he had grown up with in the Sweet Auburn neighborhood of Atlanta, which taught him that the general good of the world does prevail. But not without courageous people working in the trenches. Within that timeless teaching, a duality exists between destiny and free will. Our decisions do matter.

WHEN I ENTERED THE park at West Glacier, Montana, on the day before I found fame in The Loop parking lot, so many cars jammed the Sun Road that traffic was in gridlock. I could ride on the shoulder and cruise on past the queue. The Sun Road at the western entrance lies within the Lake McDonald valley and is surprisingly flat. I could zoom past car after car after car—*vroom! … vroom! … vroom!*—in a regular rhythm every two seconds. It was so flat that it reminded me of those vacations whereby my mother and father would drag me down to the Florida beaches every summer, and cars would be lined up in both directions on the road that paralleled the edge of the ocean. I never understood why anyone would want to sit in a car while on vacation. Wasn't vacation supposed to be, well, a *vacation*?

I had a huge grin on my face and thought, "How cool … a bicycle in the express lane. See you later!"

If you imagine the straightaway at the bottom of a roller coaster or at the bottom of a ski jump, you can see how Lake McDonald is nestled within a massive, U-shaped notch at the base of the approach to Logan Pass. In fact, on my approach to the park from the previous night's stay in a campground at Kalispell, Montana, I had only gained 259 feet in elevation over a distance of 56 miles!

It was an incredibly easy ride, and I kept anticipating, "Where are the big hills?"

Lake McDonald is a watery remnant of the enormous glacier that filled the entire valley before melting away and receding to the very high alpine areas at the end of the last ice age. At 1 mile wide and over 9 miles long, Lake McDonald is an impressive site. Its grandeur appropriately fits to scale among the high, rugged peaks of the Rocky Mountains as though a giant had been using a playset and plopped it down within the seams that run diagonally between the ridges. The Sun Road cuts across the middle of the park by splitting off of US Highway 2, which says "No, thank you," dives to the south, hugs the southern border of the park, and crosses the continental divide at Marias Pass on its 2,571-mile journey across the northern contiguous United States. The Sun Road, however, on its way to Logan Pass, runs parallel to the southeastern shoreline of Lake McDonald.

It was late in the day when I reached Avalanche Campground above the lake's upper end the day before. I had been positioning myself for this dash up to Logan Pass and down the other side in one day; due to the curfew, it was important to be right at the base of the steepest part, at the tip-top of Lake McDonald, in order to get through the pass and down the other side to the next campground before dark. Whereas the day before, when I had approached the park on US Highway 2, comfortably cruising at 12 miles an hour, today I was in the lowest gear, slowly grinding my way up the final leg to the continental divide at a paltry 2 miles an hour. At that pace, you can walk it as fast as you can peddle it. But the quick, continuous rotation on the tiny front sprocket felt to me like a meditative rhythm. I just couldn't stop without falling over. In sort of a Zen attitude, I trained my mind to accept the short, fast peddle strokes that contrasted with the *sloooww* wheel movement.

At first I thought, "Oh, my gosh! I will be here for days doing this."

When I realized that I made progress little by little, my mind became attuned, peaceful, and accepting—kind of like when a pig settles into a mud bath. The time passed and, before I knew it, the parking lot and the 10 o'clock curfew met in time and space.

The day before, at Avalanche Campground, every campsite had been full—all 87 of them. No surprise after seeing the traffic jam.

I had been in a panic though and exclaimed with despair to a ranger, "What do I do, ma'am? I can't make it anywhere else before dark. I am on a *bike*!" as though I rode on something invisible.

She pleasantly smiled. "Oh, don't worry, sweetie. We have a designated spot for nonmotorized travelers at *all* of our campgrounds."

My anxiety melted away as she pointed me in the right direction. The nonmotorized campsite was a wide-open, flat area about the size of a baseball diamond. When I pulled in, the sun was setting, and there were a dozen others there with their tents already set up in no particular pattern. We were surrounded on all sides by families who had pulled into their single, nicely cordoned slots with vehicles towing campers or who had set up big family tents around a grill or fire ring. It was as though we were like a high-density communal park inside a well-planned, medium-density subdivision.

As I was unpacking the camp stove, a communal camper came over to welcome me like he had known me my whole life. "Hey! Whatcha cookin'?"

I dug through my meager three-day food supply and said, "Oh, let's see here. Probably some rice and tuna?"

"Well, my name is Biff!" he said as he stuck his hand out.

"Biff?"

"Yeah, Biff America! From Breckenridge, Colorado."

Biff's real name was Jeff Bergeron. He was a television personality on a morning show in Breckenridge. He was known as ski country's "David Letterman." Even though I didn't know that at the time, my inclination was to play along.

"Nice to meet you, Biff America," I said as I shook his hand. "Ominous place for you to be. You know—being from ski country and staying in Avalanche Campground?" I followed it up with a raising of my eyebrows in exaggerated fashion and saying, "Get it? ... Get it? ... *Avalanche*?" I wore a half-insincere smile, and then wiped it away quickly as though nothing had happened.

He smiled and then his countenance immediately slacked. I saw the appreciation in his eyes for something I had done that should have been reserved for someone whom I knew well. My playfulness would not have set well with most strangers whom I had just met. But I had guessed with Biff that it allowed for a quick bond with someone who I didn't have much time to get to know.

By the time I started cooking, we were lit up by the fading twilight, and Biff delved right on in.

"So, where are you headed from here?"

"Well," I answered, "I will be out soon after sunrise to get a head start on the climb over the divide before the 10 o'clock curfew."

"No, I mean on farther."

I pulled out the 11 state maps that I had ripped out of a *Rand McNally* road atlas to guide me home. My plan was to do an upside-down and inverted L-shaped route that took me through the Great Lakes Region and then south at the Straits of Mackinac.

"Well, won't you be going through Traverse City, Michigan?"

"I don't know. Right now, I am just sticking to US Highway 2."

"Yeah, I am pretty sure you will be going through Traverse City. Will you deliver a note to my friend there?"

"Sure … I guess," I replied hesitantly.

Biff scribbled on an index card, including a phone number, and handed it to me. It said, "Hanz: you stud! Hope all is well. Please let this guy take a shower 'he need it' Remind him to wash his feets. Love Biff"

I FINALLY MADE IT through Logan Pass with just enough daylight left to get to the Rising Sun Campground and cooked supper in the dark. Less than two weeks earlier, I had gone through Stevens Pass in the Cascade Mountains on the first day; luckily, I was able to go from the coast, up the west side, through the pass, and down the other side in one day, and had supper cooked before dark.

Such a long first day wasn't really my plan. Though the first day of summer had passed already, it was shocking to see snow still covering much of the road and peaks around Stevens Pass. Signs read, "Remote area. Hitchhiking permitted." It was eerily quiet, no one was around, visibility was literally foggy, the sky was completely covered in clouds, and my skin had goose bumps from both the chill in the air and the utter desolation. With no direct sunlight, nothing had any color to it, and it looked as though I was watching a black-and-white film. I saw ski lift chairs hanging motionless as though everyone had suddenly abandoned the place. I expected Freddy Krueger to pop out at any moment.

But the climb to Logan Pass was quite different. As the park's website puts it, "The average gradient for this section of the ride is roughly 5.7%. To put those numbers in perspective, L'Alpe d'Huez, one of the most famous and notoriously difficult climbs in the Tour de France, climbs roughly 3,770 feet [and] has an average gradient of 7.6%. It's likely the Going-to-the-Sun Road climb would rank among the top 20 toughest climbs in the Tour de France, and would likely be a category 1 climb."[3] It made the climb to Stevens Pass seem like a small nuisance of a hill, and Stevens Pass made the population density of Logan Pass, with all its tourists, seem like a metropolis.

It took two and one-half hours to pedal 8 miles to the refuge at The Loop. I had teamed up with two other bikers at Avalanche Campground, and we locked our bikes together in the parking lot and took a hike. We got back on our bikes at 4 p.m. and rode another 8 miles to Logan Pass. The Rocky Mountains are generally shaped like an unequal, highly skewed, and tilted wedge. While Lake McDonald is snuggled in at 3,200 feet above sea level on the west side of the divide, Lake Mary is nestled in its analogous, ice-carved, U-shaped valley on the east side 1,600 feet higher. And, whereas it is a 16-mile climb from the head of Lake McDonald to Logan Pass, it is half that distance from Logan Pass to the head of Lake Mary. The uplift that created the ridge of the continental divide is made from rock that stairsteps its way up, beginning in Spokane, Washington. It is 200 miles as the crow flies from Spokane to Logan Pass, and here is how the elevations read, in feet above sea level, along US Highway 2 from west to east:

Spokane, WA: 1,843
Priest River, ID: 2,139
Kalispell, MT: 2,956
West Glacier, MT: 3,215
Logan Pass, MT: 6,647

The mountains then abruptly end 20 miles east of the continental divide. It took four days to wind my way up the western slopes after leaving Spokane. We coasted down to Rising Sun Campground, and within a rock's throw of the Great Plains, in a half hour.

SIXTEEN YEARS LATER, IN 2007, the park service would start free shuttle service to help relieve some of the congestion. In the meantime, the world continues to consume roughly 90 million barrels of oil a day, every day, 365 days a year. You can go to the US Energy Information Administration's website where they publish weekly reports that say things like, "US crude oil refinery inputs averaged 16.2 million barrels per day during the week ending August 6, 2021. ..." Unless some of the fuel made from the refinery process is put in storage, which is sometimes the case, 16.2 million barrels is how much we actually used that week. Another statement in the report says, "US crude oil imports averaged 6.4 million barrels per day last week ..."[4] If you take 16.2 minus 6.4, then you get the amount of oil that we consumed that was produced domestically. In this case, the answer would be 10.2 million barrels a day. The proportion of oil consumed in the United States and produced domestically for the week of August 6, 2021, would be 63 percent.

The debate over whether or not to drill in Alaska's Arctic National Wildlife Refuge (ANWR) pits one side against the other. One side argues that drilling is needed to increase domestic supply and drive down prices; the other side wants to protect wildlife migration corridors. Both sides probably have it wrong. What is most likely contributing to the overwhelming majority of loss of wildlife populations and creating species' extinctions is urbanization, suburbanization, and the conversion of natural habitat into farmland—not a linear, narrow, elevated pipeline corridor running across an 8-million-acre wilderness.[5]

On the other side of the debate, the number of extractable reserves in the ground in the Alaskan wilderness was estimated in 2001 to drop the price of gasoline at the pumps by about 5 cents a gallon, if that. It's because there is nothing patriotic about the location from which the oil is extracted. It still has to compete within world markets.[6]

And those markets are declining today due to the increasing economic and political pressures to end the burning of fossil. Alaskan oil already is the least profitable of almost any reserve in the world due to high extraction costs, and more and more investment firms and banks are backing away from financing Arctic oil exploration at all. British Petroleum has, in fact, walked away from their leases in Alaska because the rationale for Arctic exploration of oil has deteriorated so low.[7] Now, who knows how ANWR drilling would affect gas prices? The oil market is so volatile that in April of 2020, May futures contracts went negative. That means that if you paid for an early contract to take in a shipment of oil in May, and you decided on, say, April 21 that you're your supplies were fine and you didn't really need it, you had to pay someone to take the oil off of your hands![8]

All of this is to say that we have more in common than we think and I believe that the real policy debate should include more informative questions that would help us all like:

"How are we going to diversify our transportation network?"

"How are we going to ease the inevitable transition of jobs from one energy sector to another?"

"How are we going to allocate resources to mitigate the future effects of a warming climate in order to protect the public welfare?"

"How are we going to do all this while keeping our nation solvent?"

Going-to-the-Sun Road in the Lake McDonald Valley

Lake McDonald

2 College Park, Georgia

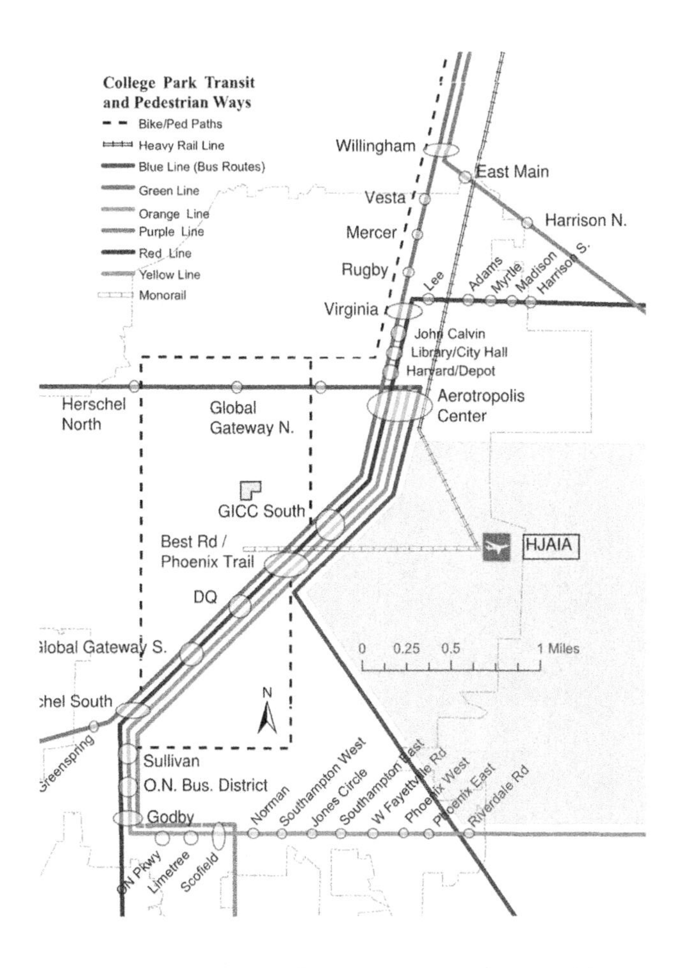

"Ninety-eight percent of U.S. commuters favor public transportation for others."—*The Onion*

STANDING AT THE BUS stop in January in short sleeves probably looked out of place. But I had spent three winters in northern Minnesota near the Canadian border, and winter in metro Atlanta just felt like a nice, cool transition out of summer. Sometimes when I get on the bus, I make my way to the rear so that I can open a window to get relief from the roaring heater. If the outside temperature on a June day were as hot as the temperature inside a Metropolitan Atlanta Rapid Transit Authority (MARTA) bus in January, I am pretty sure the air conditioning (A/C) would have been roaring instead.

The *mountains* of Georgia are a different story, however. They have four seasons, no doubt. I had also just spent a weekend in the backcountry of a remote wilderness area around a place called Tate City, camping in a foot and a half of snow. The highest ridgelines have the same ecosystem as northeastern Minnesota, whose forest canopy is dominated by spruce and fir trees and carpeted with Canada Mayflower in the summer, and its understory is covered by red-osier dogwood.

The bus driver pulled up, opened the door, looked down at me from his perch in the driver's seat, and matter-of-factly stated, "Mista Duke, you got ice water runnin' through yo veins."

The next week, I was standing at the same stop and had forgotten my transit card. I stepped on the bus and said, "Sir, I forgot my card."

With a straight face, the same bus driver, who had called me by name just a few days earlier, very formally pointed at the fare box and turned his head away and his chin up indignantly as if he didn't know me. I promptly opened my wallet, dug for some change in my pocket, and dropped the dollar bill and coins into the appropriate slots. The machine gave an approving "ding!" The bus driver nodded, and I proceeded down the aisle.

There were no open seats, and everyone was staring at me. That is the way it goes when you are the only white guy on the bus, and you are wearing a tie. As the bus made its way towards my office, the seats slowly began to clear out, and I was able to sit down the last few miles. The bus was nearly empty when I pulled the cord for the next-to-last stop on the route.

As I neared the bus's front door, the driver put out his arm and stopped me. He leaned over in his seat, pulled out his wallet from his back pocket, and handed me $1.50.

"Don't forget your card tomorrow!" he said quickly and sternly.

Bus systems are the backbone of most cities' public transportation network around the world. Bus lines are replaced with rail only when the buses start to run at full capacity. Bus comes first, then rail—not the other way around.

For example, the *Baltimore Sun* accused their own town of not being a transit city. When I visited that fair city at my step-daughter's graduation from Johns Hopkins University, I had no problem and had figured out the whole network of bus routes by the second day. For $4.25, I hopped on and off all day long and never waited longer than 7 minutes. Oh, what the heck—I just threw in a $5 bill; keep the change. In San Francisco, you ride the Muni buses to get around efficiently but not the Bay Area Rapid Transit train system. In my own home town, the complaint is that MARTA doesn't go anywhere, which isn't true; the buses actually do go everywhere, and they are part of MARTA, just like the train. In fact, some bus lines run *faster* than the train.

The bus I used to get to work left the East Point train station, went down Main Street, and then turned east towards Hapeville on Willingham Drive, at which I caught it, 10 blocks from my house in College Park. When it crossed the Clayton County line, whether or not you had paid for a monthly card, everyone had to pay a 75-cent surcharge. The City of Forest Park, in the northern part of Clayton County, had always had public transit.

But when the Atlanta metro counties voted to join MARTA in 1972, Clayton County as a whole said, "No thanks."

However, Forest Park said, "Wait a minute ... we have always had it. Just because the county says no doesn't mean that we mean no."

MARTA's viewpoint was, "Sorry, you aren't contributing anything to a $6 billion infrastructure project to upgrade to train lines and robust bus service."

An agreement was reached whereby one bus line from the East Point train station would cross the Fulton-Clayton line but not for free. Anyone getting off in Clayton had to drop 75 cents into the fare box, and anyone using a MARTA transit card to get on the bus from inside Clayton had to drop an extra 75 cents into the fare box. It was why we always had to leave the bus by the front door when in Clayton County. Normally, passengers entered at the front door by the fare box while exiting passengers left through the back door, and the bus driver had control of which doors could be opened.

Nonetheless, the "MARTA-north Clayton" fare structure arrangement was probably a fair deal for the citizens and businesses of Forest Park. While the core metro counties (Fulton and DeKalb) who *did* join the system were throwing in hundreds of millions of dollars a year into the system generated from a dedicated 1 percent sales tax, Forest Park could continue to receive service through this "cash-only" surcharge system. New or infrequent riders didn't understand this. We regulars sometimes had to explain it to them. Occasionally, a regular such as *me* got caught without the change to pay the surcharge. You didn't mind losing a quarter if you had to put in a dollar bill, but you certainly didn't want to put in a $5 bill.

I got to know a lady because we saw each other every morning, and she got off the bus a few stops before me inside the Clayton County line. One day, having neither any change nor any dollar bills, I asked if she had 75 cents that I could borrow to get off the bus.

She said, “Of course!”

A few weeks later she asked me for 75 cents.

I said, “Sure!”

Sometimes the paybacks would be delayed, but we kept an informal estimated tally in our heads. On one occasion, a “first timer” was on the bus, and he didn’t have the required 75 cents. He innocently tried to exit the back door; when it wouldn’t open, he yelled, “Back door!”

The bus driver yelled back, “Front door, please!”

The first-timer walked to the front, stared at the unopened front door, and turned around to the driver who was pointing at the fare box. This was a frustrating experience for those of us who may have been running late to work. My friend glanced at me from across the bus and pointed towards the fare box. Somehow we both knew that I was the one behind the eight ball. She was indicating that she wanted me to pay forward the 75 cents that I owed her by giving it to the “newcomer.” As the bus driver was trying to explain, I walked up and dropped three quarters in the fare box, looked at the driver, and nodded my head. The bus driver nodded in return and opened the door.

I MOVED BACK TO College Park from Minnesota in January 2010. I had owned a home in the historic district since 1998 and, while I was away in Ely, Minnesota, for those three years, flight attendants and pilots had rented rooms in the house. They were such great tenants that I usually became good friends with them and would actually stay in the house in whatever room was empty or, if necessary, on the couch, on those infrequent occasions that I visited. If you have ever taken the train out of Hartsfield-Jackson Atlanta International Airport (HJAIA), College Park is the first stop on the line to downtown Atlanta.

In Ely, I was living next to a 2-million-acre wilderness but paradoxically felt almost claustrophobic. When I made the decision to return to my home, I declared that I would never leave it again. I started working part time for Porter Alexander at Urban Solutions—a College Park city planning firm—doing a variety of environmental tasks.

Porter and I had met when I was serving on the College Park Board of Zoning Appeals. Porter was contracted by the city to, among other duties, be the liaison between the applicant requesting a zoning variance and the board. He would also make a recommendation as to whether it would be prudent to approve or deny the applicant's request. We didn't have to take his recommendation. In fact, on some occasions, I would disagree with his recommendation and sometimes would be successful in leading the charge to go against it. We never took things personally in those meetings and could be seen many times having a beer together afterwards at a local pub.

The board was made up of five members, one appointed from each councilman and one appointed by the mayor. I started on the board as the representative from Ward 3. But, after a political shake-up in my ward, the newly incumbent councilman kicked me off. I never missed a beat in the three years I served because the mayor's appointee had to leave his post at the same time and I was immediately put back on the board as his appointment at the very next meeting.

During my last meeting before leaving for Minnesota, I had opposed Porter's recommendation. He said the variance request should be denied for a reduction in parking space requirements for a new restaurant on Virginia Avenue. It was to be located about two blocks from the campus of Woodward Academy—the second largest private K-12 school in the country. The restaurant was connected by sidewalks to the school and to a large chunk of residential housing in Ward 3. Virginia Avenue had frequent, all-night bus service down Virginia Avenue, which was the sweeper between two train stations. Furthermore, it was already going to have a quarter-acre parking lot even with the variance and would have access to public parking along the side streets.

I felt the applicant had a legitimate hardship in wanting to make his restaurant larger and the parking lot smaller. It was being built up to the sidewalk, and the parking was being moved to the side and back. I thought it was a good-looking plan. The councilman from Ward 3 was not about to let me have my way and made it clear at public comment. There was much debate by everyone on the board, and I had no idea how the vote might go. After a very long discussion, I finally made the motion to allow the variance. Someone else took the bold move to second my motion so that it could go for a full vote.

The chairperson of the board said, "We have a motion and a second. Any more discussion?"

Silence.

"All in favor?"

There was a chorus of "Aye."

"All opposed?"

There was also a chorus of "Nay."

Then the chairperson declared, "The ayes have it. For the record, let it be known that Ward 2 and Ward 4 opposed."

The person in Ward 3, who had been appointed to sit in the seat that I had been kicked off of, had seconded my motion, and the chairperson broke the tie.

If you ever go to a zoning board meeting, there is honest debate about the issues. If those appointed boards are autonomous, whereby the decisions stand on their own and without oversight from the governing body, they are strong democratic institutions. There is no grand standing and, by and large, the ethics are solid, community values are met, and the right thing is almost always done. There is a chance for vibrant participatory democracy at the local level where people have a chance to feel empowered. It sustains our whole country.

Our communities, of which there are thousands, act in much the same way as a capillary does in our bodies; they are the "functional unit of society," carrying out a vital task. It's where things get done at the microscopic level and where governments are forced to set priorities in order to balance the budget. They are constantly under great pressure to accomplish this, and it can create some lively discussions.

Politics at the national level, on the other hand, feels almost lethargic by comparison. How many times in a primary debate are issues discussed, if ever? And there seems to be no intention by the federal government to *ever* balance the budget. The National Commission on Fiscal Responsibility and Reform, also known as the Simpson-Bowles plan, was released about the time I moved back to College Park and offered common-sense solutions. Under their recommendations, the tax code would be rid of $1 trillion in tax expenditures (unpaid for income deductions), tax rates would fall, and there would be a fighting chance to create a real budget that eventually leads to solvency. Any tax expenditure that we would like to keep would have to be voted on by Congress as part of the yearly budgetary process and would remain outside the realm of an unelected bureaucracy or policing by the Internal Revenue Service. You might pay more or less than what you paid under the current system. But I think it's honest and simple.[9]

The author at West Glacier, Montana

3 Hudson Bay Divide

A Pass With No Name

IMMEDIATELY AFTER SLICING BETWEEN the peaks at Logan Pass, you enter into a wedge of land inside the United States that is about half the size of Rhode Island, where the streams flow north to the Arctic Ocean. It's one of the two places in the United States where this movement happens, and I would hit both of them during my ride across the continent.

This wedge of land is formed from two ridges. One ridge, the western continental divide,[10] runs north and south through the Rocky Mountains and is the one I had transected the day before at Logan Pass. The other ridge runs roughly east and west between the Rockies and the Great Lakes, mostly in Canada but at times dipping across the border; its official name is the Laurentian Divide but informally goes by the Hudson Bay Divide. It so nearly follows the US-Canadian border that it seems the logical thing to do might have been to follow this natural ridge when establishing the political border between the two countries than to argue over an arbitrarily designated latitude line. Thus, as soon as you leave Glacier National Park, you slowly begin to climb again. I was confused and started to question myself. Various thoughts and emotions raced through my head.

I had ridden over Logan Pass the day before with Ethan and Angela, my companions from California. We had arrived at Rising Sun Campground late the night before and didn't get through cleaning up supper dishes until almost 11 o'clock. One reason was that, in fact, there was *not* a biker/hiker spot reserved in the completely full campground, and the ranger had to open an area for us that had been closed off.

We had bonded through the ordeal of a day that could not have possibly encompassed any more in 24 hours. We stopped in the town of St. Mary to get a few supplies, do a bit of laundry, and toiled up yet another hill. No one could go very fast, so we stayed close. No wind blew, no birds sang, and no cars passed. It was so quiet that I could hear my companions breathing. I was angry, weary, hungry, and elated all at the same time. Angry that I was on a hill, weary from days of uphill climbing, hungry from a chronic calorie deficit, and elated to finally be with riding companions.

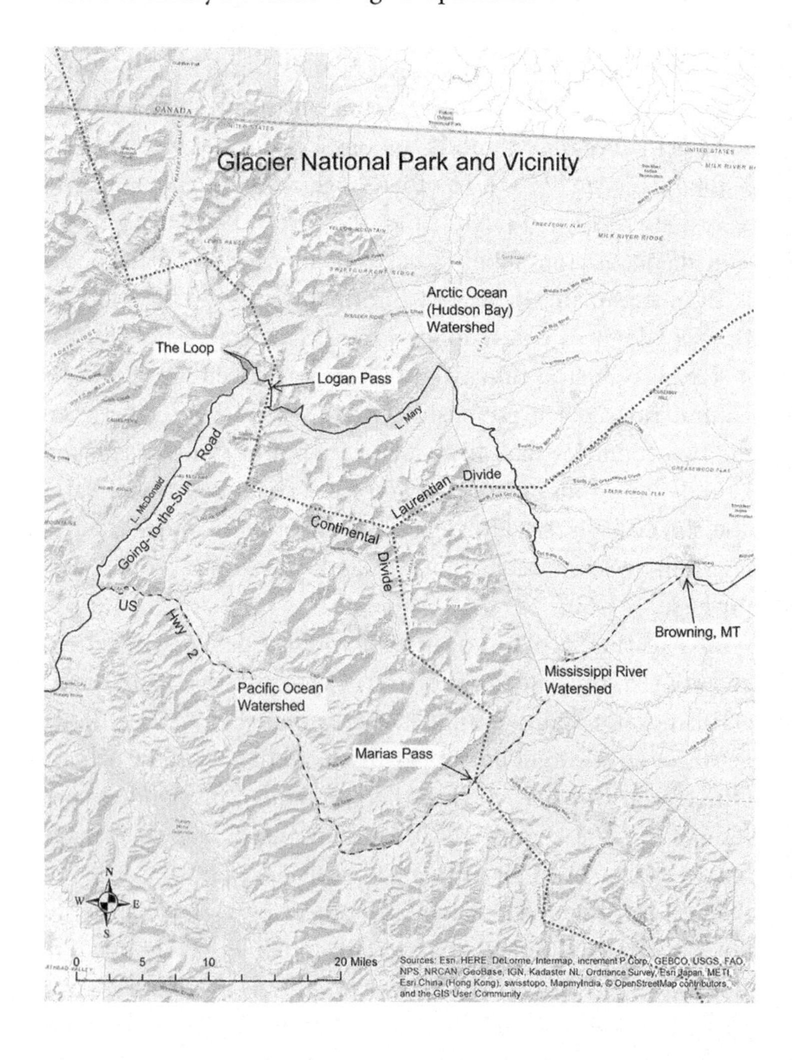

I finally yelled out in exasperation, "I thought we already left the mountains and were heading into the flat, wide-open plains!" It seemed like my words should have echoed across the barren landscape; in fact, they became absorbed and muffled by the still, empty air.

"We might be approaching the Hudson Bay Divide," Ethan offered.

I hopefully retorted, "I thought we exited the park below the divide."

Angela stopped riding, and we all pulled over to the side of the road. She got out her map and, after figuring out where we were, said somberly, "Yeah, it's another divide."

The area seemed so stale because we were within the rain shadow of the Rocky Mountains. As warm moisture is drawn off the surface of the Pacific Ocean and forms into clouds, they generally blow east. When the moisture-laden clouds hit the wall of the Rockies, they have nowhere to go but up. Cold air cannot hold water very well; as the clouds rise, the air cools, and all the moisture falls out. Therefore, the west side of Logan Pass almost looks like a cool, refreshing rain forest, and on the east side the vegetation appears stunted. Whereas big spruce and fir trees dominate the canopy in a mature alpine forest towering overhead at Logan Pass, here runt trees looked like stick figures with stubby arms almost at eye level. Whereas the groundcover at Logan Pass was lush and dense, here there were bare, brown patches of eroded soil with dry channels carved out by the infrequent bursts of rain.

To make it more confusing, I could see no mountains ahead of me. We were just facing a dull, paved incline in front of us. The air was at a complete standstill and I felt, although dry, uncomfortably stuffy.

I wondered, "Are we *really* climbing just one, big, isolated hill out in the middle of nowhere?"

This was the exact spot where the Laurentian Divide made a quick dive across the Canadian border and into the United States where it continued to a "T" into the continental divide at a summit inside the park called Triple Divide Peak. The T forms three wedges of a pie: the western wedge drains to the Pacific Ocean, the eastern wedge drains to the Mississippi River, and the smaller northern wedge drains to the Arctic Ocean.

We had fallen inside the northern wedge after crossing Logan Pass the day before. We had dropped into a little nook that happened to empty towards the top of the globe and were climbing our way up and over in order to get back into the Mississippi drainage basin. I understood why the major transportation corridor had been built through Marias Pass: Why go over two passes when you can accomplish the same thing with one?

When I turned out of the park on Highway 89 towards Browning, Montana, I had thought we had smooth sailing on flat, wide-open roads for the next 900 miles. We found ourselves slowly grinding uphill at 2 miles an hour ... again. We were climbing back up to the "US" side of the Laurentian Divide and, after 5 miles, we finally reached the pass that didn't even have a name!

We all got off of our bicycles and stood in the middle of the desolate road, bellowing the poetry from a '70s band named America that could best describe the contrasting landscape over the past two days. One day we were in a lush forest; the next we were in a desert.

Angela suggested that we stay down a dirt road that dives back into the park from its eastern edge to a place called Cut Bank Campground. Not surprisingly, that campground was full too. Except in this remote part of the park, there was no ranger to appeal to. I was tired of muscling the heavy bicycle around on the 5-mile-long entrance road that sometimes made the tires sink in. It took nearly an hour to get to the campground once we left the pavement. I wanted to relax and have one of the beers we had bought from the grocery store in St. Mary.

"Now what do we do?" I asked. "Go back to the main road?"

As Ethan was contemplating the map, he said, "Well it looks like …"

"Hey, guys," Angela interrupted. "There is a fellow over there who is waving at us."

A man with black hair and brown skin had his arms in the air and was staring us down. "You guys looking for a place to pull in?" he yelled across the rode with his hands cupped around his mouth.

He was the father of a Native American family from the Blackfeet Indian Reservation. He was offering his spot and wanted us to have the leftovers from their day's picnic before they went home. We ate so much of their food while we listened to his stories that we didn't even bother to cook supper that night.

The next day, we stopped at a grocery store in Browning, a quiet town on the edge of the prairies in the middle of the Blackfeet Reservation. I saw subtle signs of utter abject poverty: rusted-out cars and rundown mobile homes; some were burned out. Some of the same tribe of people who had fed us the day before were asking us for money and/or food while we packed our groceries.

I felt embarrassed that I had taken food from some of these same people the night before and now they were asking me for help. How could I not buy food for those who asked? We went back in the store and nearly spent as much on others as we did on our own supplies. We ate lunch near the Museum of the Plains Indian and afterwards Angela and Ethan began the journey back to California by turning towards Marias Pass. As I pedaled away, I turned to catch one last glimpse of them, waved goodbye, and have never seen them since.

The rain shadow of the Cascades has a similar look and feel as the dry climate east of the Rockies.

4 The Great Plains

"Ain't this place a geographical oddity—two weeks from everywhere!"—Ulysses Everett McGill, *O Brother, Where Art Thou?*

THE LAND TO THE east of Glacier National Park is called the "high" plains for good reason. At Browning, Montana, headquarters of the Blackfeet Indian Reservation and where I reconnected with US Highway 2, the elevation is 4,377 feet. Browning is higher than many peaks along the Appalachian Trail. But, in Browning, it is as flat as a tabletop. It's as though the flat ground has been lifted up along the base of a wall known as the eastern side of the Rockies.

There was a blue sky on the morning I left the park on US Highway 89, leading to Browning, with my back to the mountains. It occurred to me to take a moment and see what was behind me. I pulled over, got off the bicycle, and slowly turned around. I looked due west and 11 miles as the crow flies from the border of the park with my mouth slightly gaping open. The stark contrast between the bluish 10,000-foot peaks of the Rocky Mountains, which were illuminated by the rising sun behind me, and the dry, brown, sandy, wide-open, nearly treeless landscape in front of me was extraordinarily impressive. The two-lane highway was just a meaningless, almost unnoticeable, thin linear piece of slithering, black asphalt, which ran off and disappeared into the distance in the middle of a wide frame that extended beyond the limits of my peripheral vision. I had to sweep my head from side to side as if visualizing a panorama to take it all in.

"Bye-bye, mountains. I will miss you."

Five days and 422 miles later, I pulled into Culbertson, Montana. JJ, my friend from college, worked on a beef farm in the vicinity as a seasonal hand. Culbertson is located in Roosevelt County, population 10,425. It is dry ranch country and, in this area of the plains, though vistas still seem to extend to the horizon, there is some rolling undulation that begins to provide some variety to the flat and wide-open landscape. Population density is 4.4 people per square mile. In 1993,

Dayton Duncan wrote a book called *Miles From Nowhere*, which debunks the myth that the American frontier had been declared over in 1890. He traveled to several counties west of the Mississippi River and interviewed people who still live in places that have less than two people per square mile, the definition of a frontier that had been arbitrarily set by the US Census Bureau.

Roosevelt County has over twice the density of the places Duncan visited, and *it* was pretty desolate. The only reason that Roosevelt County didn't fall into the frontier category was because of the lifeblood of the railroad, US Highway 2, and the Missouri River. There were 23 counties in Montana that were classified as frontier when I rode through the state in 1991. Most of them are east of the Rockies, and I hit four of them along the Hi-Line, the common name for the major transportation corridor running just south of the Canadian border from Spokane to Minnesota. Their total population has dropped another 7 percent. Agriculture—wheat, barley, and cattle—comprises almost the entire economy. Little did I know that I would soon experience it firsthand and become a part of it.

Even though it is several hundred miles from Glacier National Park to Culbertson, they still lie within the rain shadow and usually receive only about 12 inches of rain per year. I had slowly dropped 2,400 feet in elevation since leaving Browning but had never noticed it. It was such a flat and open landscape that I saw the white sides of buildings in most towns of Montana 3 miles before I arrived. Once in town, I could fill my water bottles at the local park and get some shade because it was the only place where trees had been planted. The wind had been in my face, and at times I had been counting pedal strokes to break up the monotony.

I found a pay phone, deposited one single dime, and dialed JJ's number. Someone said that he would have to call back. I gave them the number on the pay phone and waited. About 10 minutes later, the phone rang.

"JJ?" I asked unconfidently.

"Yeah, it's me. Where are you?"

"In town ... Culbertson. You can't miss me. How about meeting me at The Coach Lounge?"

"Be there in about an hour," he replied.

"Okay. I will be waiting."

The population of Culbertson is 779. Like most towns across the great plains of Montana north of the Missouri River, US Highway 2 is "Main Street" and usually runs less than a mile long. In Culbertson, US Highway 2 is called 6th Street, though. Back in Chester, it was called Washington Avenue. In Havre, Chinook, Malta, and Glasgow, it was 1st Street. In Harlem, it was Central Avenue. And, at Wolf Point—the county seat of Roosevelt County, which has the largest community of Native Americans on the Fort Peck Indian Reservation—it's just US Highway 2. The total population of those eight towns, which is widely spaced out like beads on a string, is about half of the number of people at a college football game on a typical fall Saturday.[11] It is not hard to be found.

All of these towns were originally built as farming settlements along the Great Northern Railway, and US Highway 2 came later as a parallel thoroughfare. It is thought that this railway inspired the fictional Taggart Transcontinental Railroad by the author Ayn Rand.[12] Against the backdrop of the wide-open plains of Montana and the endless sky, the train cars look like block toys.

One day, a freight train tapped its horn at me in a gesture of "hello." This brief human-to-human connection, though in the distance and from someone whom I couldn't even see, inspired more emotional bond inside me on this lonely stretch of highway than any of the characters ever could in *Atlas Shrugged*. There are silos to store grain in every town along the railroad line where freight cars empty them and haul the cash crops away to coastal ports. Today, Amtrak shares the line transporting people between the Midwest and Seattle, with stops on each side of Glacier National Park. Its route is aptly called the Empire Builder. Rand's protagonist, Dagny Taggart, would have liked that.

RAIL TRAVEL INCLUDING BOTH passenger and freight trains accounts for less than 1 percent of all greenhouse-gas emissions in the United States. On the other hand, vehicles—from cars to delivery trucks but not including 18-wheeler freight trucks—are one of the top two biggest shares of US greenhouse-gas inventories, and they create *almost six times* more than all the jet airplanes taking off every day.[13] Here is a proportional pie chart I created from US Environmental Protection Agency sources, depicting where greenhouse gases are derived so that you can see the big chunks:

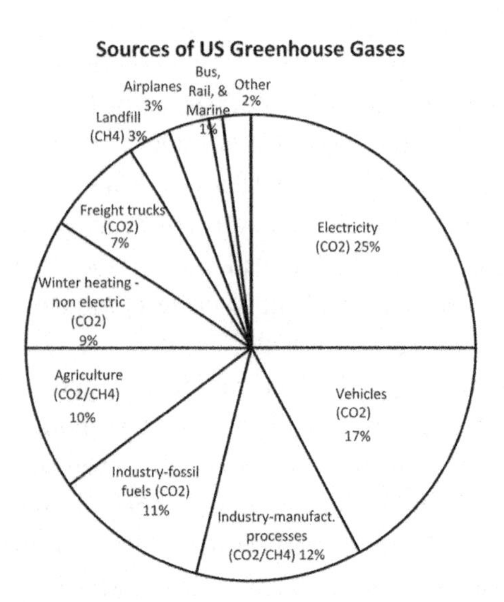

Let's face it. Everybody loves cheap gas. What if we published and discussed electric rates in the news like we do gas prices? There is about 34 kWh equivalent of energy in a gallon of gas, and the Chevy Bolt gets about 4 miles per kWh. Let's say the average kWh costs 18 cents in North America, and gas is running at $3 a gallon. That means the Chevy Bolt is refueling at about a dollar "a gallon."[14] Electricity is so cheap in some large cities that it is fairly easy to find free electricity. A colleague of mine has driven on free fuel from Atlanta to New York City! Strangely enough, a trip from Atlanta to Nashville and back can be done purely at Harley-Davidson free DC fast-charging stations. But the gas engine will always be 0 percent carbon-free while the electric grid is getting cleaner every day. Plus one big power plant, whether or not it is fossil, is way more efficient than millions of tiny engines with personal tailpipes. It varies by region, but the average EV emissions as gasoline miles per gallon (MPG) equivalent is 93 MPG in the US.[15] In Canada, the average is well over 100, with over half of provinces above 300 MPG![16]

EV Emissions as Gasoline MPG Equivalent

Average EV, 2021*

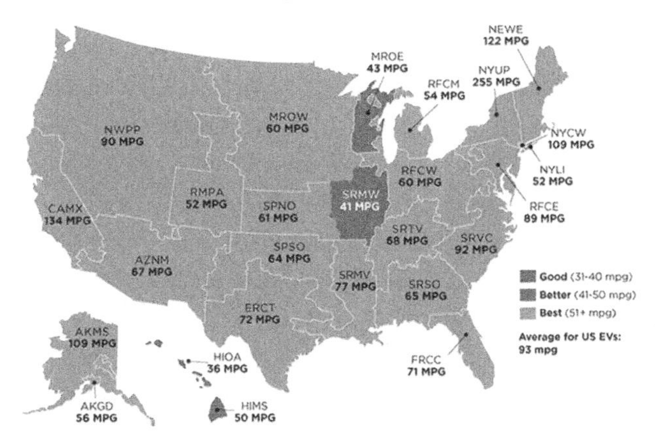

* based on 2019 reported electricity generation emissions

© Union of Concerned Scientists

This is an important lesson because, in early 2015, Christophe McGlade and Paul Ekins, in the journal *Nature*, implied that not only will we *not* run out of gas (at least not in our lifetime) but that we should probably leave at least a third of the world's oil supply just lying there in the ground. Just let it sit![17]

Ecological principle #1 is that the energy I use to propel my bicycle and the energy used to run a gasoline engine both began as solar radiation. It gets converted from one form to another and, along the way, all of it is eventually lost as heat, which is released back into outer space. The difference between me and the car is that I am using "real-time" energy. I ate some beans and rice that were grown within the past year and converted the stored energy to pedal my bicycle. As I pedaled, everything was used up and some was converted into heat, which went into the air and (mostly) escaped our planet into outer space through the atmosphere.

If you heat a house with a wood-burning stove, you are also using real-time energy. The tree had been storing energy from when the sun shone on it during the past 20 or 30 years before it got cut down. You burn the stored energy in your stove during the cold winter and, at the same time, another tree somewhere is taking in CO_2 and storing away more energy. If this procedure is managed correctly, and we don't denude the trees too much so they can't replace themselves, it works out pretty well. However, burning oil releases energy that had been stored by what were plants, which had been buried, compressed, and sitting dormant within the crust of the earth for millions of years.

Ecological principle #2 is that no molecules—such as carbon, oxygen, or water—can ever go off into outer space. (Nor can any enter.) Theoretically, not one single atom. One thing that allows life to exist on the planet is the fact that the atmosphere is like a dome that holds everything in and keeps out harmful ultraviolet radiation. All of these different molecules can hit the dome's underside but will bounce back towards the earth (i.e., terra firma proper as the atmosphere is an integral part of Earth). They can get redistributed and moved around the planet, but they can never leave.

We can actually measure all these molecules and track where they currently reside. For carbon, about 1,000 billion tons have dissolved into the surface waters of the oceans; about 1,500 billion tons are in microbes and decaying plant matter within the soil; the atmosphere holds about 800 billion tons; and there are about 600 billion tons in plants. All these reservoirs of carbon are fluid and dynamic. Carbon readily moves around these four places on a daily basis.

Imagine all the credit-card machines at stores and banks that are moving cash around every day in order to operate our economy. But what if someone added extra money into the economic system every day? Inflation would eventually get out of control. When we are not using carbon in real time (i.e., when we are burning mined fossil), we are adding to these four main carbon "players." And so we have figured out that you can budget carbon just like you can budget your checkbook.

The average annual global temperature before the Industrial Revolution was less than 57 degrees F. In 2016, the hottest year on record, the average annual global temperature was 59 degrees F. By comparison, Mars' atmosphere is flimsy (1/100th as thick as Earth's), and it can hold in only enough heat to keep the surface at an average temperature of *negative* 57 degrees F. Venus, on the other hand, has a thick atmosphere made of mostly CO_2 and has a near-constant temperature of 878 degrees F. Our moon has no atmosphere and has a hard time even regulating heat, and fluctuates from a scorching 257 degrees F during the day to an almost unbelievable –258 degrees F at night.[18] We have about 900 billion tons of carbon left in our budget that we can burn from fossil fuels to keep the average annual global temperature below 61 degrees F.[19] That is about 25 years' worth of carbon that we are emitting at today's rate (36 billion tons annually). We could continue burning at today's rate until the year 2046, and then go to zero in year 2047. Not likely. Better start reducing emissions today.

For Earth, what does a 1-degree change mean? During the last ice age, which created the glaciers that carved out the Lake McDonald Valley, the average annual global temperature was 48 degrees F, about 9 degrees colder than today. Minnesota was covered in a sheet of ice 1 mile thick. Some of the ice still exists, but it has retreated northward and now sits in northern Canada. There were still four seasons (outside the tropics), just like today. However, so much precipitation fell over centuries and never returned to the oceans because it remained frozen, and ocean levels dropped so much that the Bering Sea became a wide plain covered in snow in the winter and was covered in grass in the summer. Think of a cold, Serengeti ecosystem between Alaska and Russia where today there is a 53-mile-wide, 130-foot-deep gap covered by salt water. To keep the planet cool, we will have to leave a third of the world's oil reserves in the ground, keep half of the natural gas reserves buried, and leave 80 percent of the coal reserves intact.

I WENT TO THE RANCH with JJ that night. I had no idea where we were. We drove somewhere towards the Missouri River, which I never saw, and I don't remember seeing another house or even another car the whole way. Every once in a while, we would slow down to drive across cattle guards that would make the tires rumble. Cattle guards were created to allow a break in the fence where public roads transected the fence line and no gate exists. They look like stormwater drains with parallel slats that sit level with the grade of the road with the idea that cattle won't walk across them. I remember straining to see out the windshield as the headlights illuminated the dust in the air ahead of us. The small piece of road inside the beam of the headlights was framed in pitch black darkness.

When we finally arrived, JJ said, "Oh, yeah. We are moving cattle to fall pastureland tomorrow."

"In July?"

"Yeah. Winter here starts sometime in October. Want to help, or do you want to hang here?"

"I'm with you."

"We will put you on a horse to help push the herd along."

"Cool!"

"Breakfast is at sunrise."

They put me on a chestnut Appaloosa-Thoroughbred mixed gelding named Clyde. We moved 250 cows and their calves for 12 miles. About half the route took us across wide-open rangeland, and the other half went literally down the middle of a dirt county road bordered on both sides by fencing. Clyde mostly walked the whole way, but every once in a while on the open rangeland, a calf would mindlessly stray off to the side, and I had to nudge Clyde into a light trot to head the calf off from wandering God-knows-where and redirect its course back towards the herd. I could see the dumbfounded calf turn its eyes and head slightly to the side, as if to say, "Where did you come from?" and then it would bolt back into the herd, usually with its back feet kicked up into the air as it pushed off.

The ranch spanned across 20,000 acres that included the bottomland floodplain of the Missouri River, where big cottonwood trees grow. It also included the rolling plains of the high range. Sometimes in between the bottomlands and high range, you can find the badlands that are a unique feature to this part of the continent. The badlands consist of patches of exposed rock. If you ever wondered what it would be like to stand on the moon, this is as close as it might feel. The ranch was totally self-contained. JJ told me that, when it was time for him to leave, they would load up the wheat truck with grain and drive it into Culbertson to be sold. This marketing task would generate about $2,500 cash to pay JJ at summer's end.

The farm grew wheat, barley, and oats as commodity crops to be sold. The farm also had its own feedlot, which was key in making it totally self-sufficient. Most operations that raise beef calves ship them off when they are weaned at three to four months old to Iowa or Kansas or somewhere else in the Midwest where either rainfall or an underground aquifer provides more plentiful water compared to Montana and where they can grow enormous quantities of corn. The weaned calves eat corn for about a year and are then shipped a second time to the packing house.

The calves raised on JJ's ranch, however, were shipped only once. The ranch grew corn and safflower but had the advantage of being able to draw irrigation water from the Missouri River. Once harvested, these crops were placed in earthen fermentation pits and made into cattle feed called *silage*—probably a similar process to making sauerkraut from cabbage—and thereby preserving it for the coming winter. Instead of shipping calves to a feedlot in the Midwest, the silage was fed to the calves once they were weaned. The commodity grains, also referred to as "dry crops" and "cash crops," could then utilize the highland because they would normally thrive in the relatively dry climate without irrigation. There was a small plot of irrigated land around the house, which grew vegetables to be canned for the winter. Probably about 50 percent of the ranch's total food demand was met by everything produced onsite; the rest was bought in town with some of the beef- and dry-crop revenues.

The cows could satisfactorily utilize the grasses growing on the open range during the frost-free months of the year for all their caloric needs. East of the Mississippi River, one cow can maintain itself and produce milk for a growing calf on 1 or 2 acres of grassland. In Montana, due to the parched conditions, it takes about 60 acres per cow.

Besides growing grains and silage, some of the ranch was also used to produce hay for the cows during the winter months. By the time winter came, a cow's caloric requirement was reduced because the calf had been weened. But those caloric requirements were replaced with the energy needed to survive a cold Montana winter. During the summer, when grass was abundant, she needed to produce milk, which itself is four times more caloric-demanding than pregnancy, and put away fat stores.

Montana: Big Sky Country

When we finally reached our destination and had turned those cows out on their new green field, they immediately put their heads down and slowly walked as they devoured the blades of green grass. Not only did the green grass provide calories, it was succulent enough to provide a bit of water stored in the grass blades. They say that a cow on the western rangeland has to walk 1 mile an hour all day long to stay alive. Before October, the calves would be rounded up and sent to the feedlot, bulls would be released, and a new crop of calves would hopefully be born the following spring.

IN CULBERTSON, I CALLED Caroline and, when she answered, she gave me a mock scolding: "Where have you been? Have you called your mom yet?"

"Yes. Mom knows I am still alive. You know, it is wide open out here, and towns are few and far between. Hey, listen. Can you ship my tent to the post office in Williston, North Dakota, care of general delivery?"

General delivery meant that you walked into the post office, showed your identification, and they gave you your package. I had been camping under a rain fly. It kept me dry, but it didn't keep the mosquitos out, and they were getting worse by the day. If Caroline got my tent in the mail soon, it would reach Williston before I got there. If not, I would have to wait or have it forwarded to Minot, which was 125 miles even farther down the road.

"That's it? You want me to ship your tent?" she asked incredulously.

"Yes. please. Mom and dad are on vacation. Do you have a key to the house?"

"Actually, I don't."

"I will let the neighbors know that you are going to pick up a key," I explained.

"Okie dokie. When do you need it?"

"Well, if you don't get it in the mail pretty quickly, I could possibly pass it up. Can you send it to Williston, North Dakota, tomorrow?" Williston was the first major town across the Montana border.

She gave a disappointed "sure," as if it were the only reason I called. There was silence on the phone, and then Caroline sounded hopeful: "By the way, I was with your friends Carlos and Brodie last week!"

"Oh, that's nice."

"They miss you!"

"Tell them hello for me."

Another disappointed "okay" came from the other end, and then a hopeful, "Hey … you *are* coming back to school, aren't you?"

"Why do you ask?"

She seemed to be changing subjects and cheerfully said, "You know I spoke with Professor Phelps when I got back home?"

"And?"

"Well … the word on the street is that you aren't returning to school."

Professor Phelps was the most gregarious teacher we had. He knew me quite well, and sometimes purposefully and deliberately called me out in class. But I had never betrayed my deep, personal feelings with him—especially not anything about the misgivings I had about my choice of academic path.

"The rumor is that you might have competing interests and that you won't be here in the fall, at least not in any of our classes," she blurted out quickly.

I quickly changed the subject: "Will call again soon. Please get the tent in the mail as quickly as you can. Thanks, and have a good week!"

5 Newcastle, England

"There is a light at the end of the tunnel. No, wait. That was just an oncoming train."

CAROLINE AND MY RELATIONSHIP would eventually fail, but I did graduate from vet school two years later and started working in the stockyards as a contract veterinary epidemiologist for the US Department of Agriculture's Federal Brucellosis Eradication Program. I was usually covered in cow manure at the end of the day, but it was my chance to work in the public health arena, and my time with JJ on that Montana ranch may have helped seal the deal on my decision to return to school.

I was really what you might call a "disease surveillance specialist" and was paid $1.50 for every blood sample I drew. The federal government was intent on wiping out bovine brucellosis for good. That is a hard thing to do, but they came close before the program costs were transferred to the farmer due to funding cuts. If they didn't entirely eradicate the disease in the United States, they made a good effort. The problem is that the disease could pop up in wildlife herds of bison out west, especially around Yellowstone National Park, which might come in contact with free-ranging domesticated cattle. And then the cycle might continue due to this wildlife reservoir of infection.

Working the stockyards where cattle are bought and sold was the most fun job I have ever had. Sale day was a whirlwind of activity. We had less than 60 seconds to draw the sample before the next cow ran into the chute. On hot summer days, I soaked in sweat and lost a few pounds. I felt like I was part of a large cause. There was a sense of mission—as close as I would ever come to what might be akin to how it was for the World War II generation. Some days we tested over 900 cattle; other days, when beef prices fell due to drought or from a public relations issue, such as mad cow disease, the number of cattle coming through might fall below 100.

When the program ran dry and all federal funds were cut for contract epidemiologists (I suppose $1.50 per blood sample was too high to survey for a disease that had been nearly wiped out), the US Department of Agriculture sent me to the United Kingdom to help fight the outbreak of a vesicular disease—think cold sores and fatal in young animals—that was wreaking havoc on the food supply. The disease is a virus that was spreading through the sheep herds of Great Britain like wildfire. In fact, a similar operational infrastructure established by the US Forest Service in containing wildfires was used by the epidemiologists overseeing the enormous number of resources that were pouring into Great Britain. There was a sense of mission, much like the stockyard program in the United States. The difference is that there was a greater sense of urgency.

Brucellosis is a slow-moving, blood-borne, bacterial disease, mostly among young heifers. Here, the virus could even be spread from a vehicle driving from one contaminated farm to another, and whole farms were being eradicated. People were losing their livelihoods; to make matters worse, the people were basically being quarantined on their own farms. Sheep dogs had "lost their job" and didn't know what to do with themselves. At times, I felt more like a grievance counselor than a surveillance specialist because I often became the only social contact they had due to the quarantine status imposed upon them.

As soon as I arrived, I was put on what they call "hot patrols," which are done within 3 kilometers of an infected premise to check for spread to nearby farms. "Surveillance" is done outside the 3-kilometer circle of an infected premise but inside a 10-kilometer circle. I had flown into London on August 23rd (2001)—six months after the epidemic had begun.

I had a debriefing the next day and was then sent to Newcastle upon Tyne on the following morning's train.

On the first day in Newcastle, I listened to lectures on protocols all morning; by the first afternoon, I had my vehicle stocked with supplies for the next day's farm inspections. Things moved fast, and conditions were constantly changing.

Each day, I reported to "Allocations" at the Disease Emergency Control Center—a.k.a. The Deck—that provided oversight and coordination for all the data coming in from the field. There were trailers for extra office space for all the different personnel. Some had come from places like Australia and Germany and had been there since the disease broke in February. Those "veterans" had been moved up to team leaders. New maps, with overlapping hot patrol and surveillance circles due to multiple infected sites, were created and posted on the wall every morning.

A paper sign on the wall had this handwritten notice: "Hot Patrols clarify instructions for your work with team leaders before going out. ALL COMPLETED PAPERWORK MUST BE RETURNED TO TEAM LEADERS (NOT ALLOCATIONS) FOR AUDITING."

My days usually revolved around the vicinity of the Town of Haydon Bridge, about a 40-minute drive to the west of Newcastle on the A69. Haydon Bridge, which is named after the existing pedestrian bridge across a fork of the Tyne River, is near the ruins of Hadrian's Wall. Even the mighty Roman Empire couldn't handle the feisty Scottish. To keep them out, they built the wall across the island of what is now present-day Great Britain along the shortest distance possible: between the North Sea to the west and the Irish Sea to the east. It was the most northern extent the Roman Empire would ever reach.

On my daily duties, I came across remnants of this wall, which was built about 100 years after the birth of Christ by the Roman emperor after which the wall is named. There are still archeological remains of fortifications in one or two spots along its 73-mile stretch. However, at least in the area of Haydon Bridge, much of it had been pillaged to build Langley Castle during the 14th century; over the years, farmers had slowly picked it clean to cordon pastureland for their sheep herds. The contemporary political boundary between England and Scotland still roughly follows the old wall.

The best food I could find was in Haydon Bridge at a pub called the General Havelock. I got tired of the standard fish and chips at many restaurants. The General Havelock had a variety of choices from the European menu. Some days, the special would be Toulouse sausage with a bean sauce; other days, it might be pasta with tomato and basil lamb sauce.

The British language was another thing to adjust to. Sometimes someone would hang up the phone at The Deck and say, "He must have been Posh British," meaning "high class." The most famous saying (available on refrigerator magnets) is the announcement on the London Tube: "Mind the gap," implicating you to be careful of the narrow space between the platform and the train. A sign on the Newcastle subway might say, "Always have your ticket with you *whilst* traveling." Or a sign might say, "Enter and *alight* quickly," which I assumed meant to get somewhere and hold on! The exit sign had an arrow and the words, "Way Out." Someone might say, "It'll cost you about 50 quid" (one quid is one British pound). And to "queue up" means to stand in line. People would respond quicker if you asked where the "loo" was instead of the "bathroom."

One morning, a farmer named Mr. Cunningham called me and said, "Can you be here at half eight," which meant 8:30

a.m. It was my second visit to Mr. Cunningham's farm. When I arrived, he said, "You know, Miss Dot here can do the work for us."

Dot was his Border Collie. I had seen these dogs work before, and they were astounding. It meant that I wouldn't have to even cross the gate—no biosecurity risk and no scrub down, just look over the 4-foot stone wall from the roadside. The first visit required a hands-on look and examination of a sample of the herd that would give us a statistical probability of whether or not the disease was present. If there were 50 sheep in the herd, then all 50 would have to be examined. But if there were 500 sheep in the herd, according to the statistical calculations, I would only have to examine 93. On the next visit, it was possible to just observe the herd's behavior. If there were any that seemed to be depressed or were limping, then the whole examination would have to be completed again.

As I looked over the wall, I saw no sheep anywhere in the vicinity whatsoever and asked Mr. Cunningham, "You mean Dot can corral them into this area beside the road?"

"For sure."

"How is she going to get them through that small opening over there that connects to the wide-open back fields?"

"She will slow down and carefully tend to them."

"Sounds good to me!"

Mr. Cunningham stood at the wall with me, whistled, and off Dot disappeared into the back fields. There was a hedge of trees in front of a gentle knoll beyond the corral area that blocked our view of where Dot had gone. The corral area was an open, grassy area about the size of a soccer field. The hedge of trees slowly bent towards the wall farther down the line of stone to eventually contain the field on three sides. Mr. Cunningham had placed a small, wire fence from the wall where we were standing straight to the hedge of trees to complete the containment of the field on four sides.

About eight minutes later, a few sheep poked their heads through the small opening into the field. We could now see Dot crouched down and moving from side to side, according to the different pitches of Mr. Cunningham's whistles. It was as though Mr. Cunningham were moving Dot with an invisible hand. Dot's reactions to the whistles were instantaneous. I held my breath—one wrong move and they would scatter in chaos. At Dot's gentle persuasion, they slowly and patiently moved in a tight group, as one body with many legs. It was one of the most extraordinary events I have ever witnessed. As they walked across me from left to right, I singled out individual animals—like a wolf would see which one is vulnerable. My job was to detect any hint of drooping heads, salivation, or lameness.

Mr. Cunningham told Dot, "That'll do."

She stopped and fell to her chest with her head up and front legs forward. This position allowed them to relax and be in a less heightened state, so that we could detect any abnormalities. Mr. Cunningham and I agreed that the herd looked healthy and, with the extraordinary help from Dot, I was able to be ahead of schedule for the day's events.

One day out in the field, I was about to walk on a farmer's premise when I got a call from Phil Kophaus in Allocations. It was the first time I had ever used a mobile phone.

"Hey, John. This is Phil. Have you scrubbed down yet?"

"Yes, sir."

"Have you entered onto the premise?"

"Not yet."

"Good. You are still clean and will be allowed to continue to patrol. Please go to Hunter Oak Farm and get a report in by this evening."

"Yes, sir. Will do."

Phil was my team leader. He was a dairy veterinarian from New Zealand and had been working on the epidemic since nearly the beginning. He must have gotten information that the farm I was about to enter might be infected already due to a neighboring infection. After I had been there a week, he had okayed me to have a day off. It couldn't have come at a better time. The morning before my off day, I had been on an infected farm and, even though we scrubbed our clothes down before and after leaving any farm, whether infected or not, you were strictly prohibited from visiting another farm for 72 hours after being on an infected premise.

If I had not had a day off, I would have been moved from hot patrol to "cull assignment" and may have never gotten back to patrolling clean farms. In some eyes at Allocations, I was just a worker bee to be used any way they saw fit. But to Phil, I could sit in the office and do paperwork for a day during my 72-hour quarantine because he valued me more as a surveillance specialist than anything else. Thankfully, I would be able to continue patrols with the farmers whom I had already established a relationship. A sign had been put up at headquarters that summed up the chaotic atmosphere: "There is a light at the end of the tunnel. No, wait. That was just an oncoming train."

I had been in London with three of my American colleagues from the beginning. Amadi and Gavin were from Colorado, and Anna was from rural South Georgia. This was Amadi's second tour, and he helped us navigate the bureaucracy and politics of the Department for Environment, Food, and Rural Affairs—an agency beneath the Ministry of Agriculture, Fisheries, and Food, and the United Kingdom's equivalent to the US Department of Agriculture.

We were brought into the epidemic near the end, and there was a lot of frustration in getting the disease completely eradicated. Rumors were flying about where the disease had started six months earlier, if it was a hoax, and why the higher ups weren't allowing vaccination. We knew it wasn't a hoax—there had been virus isolation in the laboratory—but the answers to the other questions were complex. Nonetheless, people were growing weary.

On the afternoon of September 11, 2001, Gavin called me to ask if I had heard the news. The United Kingdom is seven hours ahead of the Eastern Time Zone.

"I have been away from the phone all morning, inspecting sheep at Brokenheugh," I replied.

Gavin said, "Well, turn on the TV when you get back to the hotel. Something with terrorists has happened back home."

I thought, "Just another silly scare."

Little did I know the horrific events that were unfolding. The Brokenheugh Farm overlooked Haydon Bridge and was right in the middle of a hot spot. I had a double weight of sadness. The next day, the farmers whom I had come to know were telling me how sorry they felt for America's loss and yet they were losing everything about their livelihoods.

I consoled them by responding, "Thank you. But your suffering is no less, just under different circumstances."

I was now juggling about seven different clean farms for follow-up visits on two- to three-day intervals to recheck for presence of disease. On September 16, 2001, I did the final inspection of Mr. Cunningham's small farm. He had been caught inside the 3-kilometer hot zone, but it had been 21 days since the initial visit. I was responsible for his farm from start to finish, and he was now completely cleared of disease. We contemplated for about an hour while looking over the stone wall at his sheep.

Mr. Oaks of Hunter Oak Farm drove up and joined us, and started talking about American beer: "I've seen lemonade darker."

Mr. Cunningham chimed in, "Hard to get pissed, isn't it?" ("Pissed" meant "drunk.")

Whereby Mr. Oaks responded, "I guess you can. It just takes a bit o' work."

The last confirmed case occurred on September 30, 2001.[20] The United States was pulling us out and not sending anyone else in due to the events of September 11th. My tour would officially be over on the first day that planes resumed flights across the Atlantic. I got on a 747 to Washington, DC, on September 22, 2001. The plane was enormous yet so empty that it seemed like you could have played catch with a baseball. I laid the armrests down on the seven or eight seats in the middle row and took a nap.

BY THE TIME I reached North Dakota, it was impossible to eat breakfast in camp due to the increasing mosquito population, and the tent couldn't have arrived at a better time. There was a consistent group of about 10 bicyclers heading east. Some were headed to the University of Michigan; one group was going to Montreal; others had itineraries that took them to distant places elsewhere. I only saw them at night; there were few places to camp in the endless grasslands, and we all usually found the same place in a small town about 75 miles farther down the road, give or take 5 or 10 miles, from the previous night.

Each morning, I would pack as fast as possible once I left the safe zone from inside my tent's netting. The swarms of mosquitoes, which consisted of hundreds if not thousands of individuals, would literally—and I mean *literally*—drive a person mad. Packing was an event to be done in as few seconds as possible. I didn't even think about trying to boil water on the small camp stove that I hauled around to make a cup of coffee at the picnic table. We rode to the nearest café in the nearest town to get coffee and breakfast inside the safe haven of a building. The breeze created from propelling the bicycle was enough to keep the swarms away.

One morning in the town of Lakota, we stopped at a café that was adjoined to a Chevrolet dealership. Our waitress worked the whole place, including the counter where about eight men in overalls were gathered and who appeared like they had sat in that same exact spot every morning for decades. The waitress wore a white apron as she hustled around on her chubby, short legs and appeared to still have the same hairstyle from when she was a teenager in the 1950s. You had to walk across the showroom floor to get to the bathroom. We looked comically juxtapositioned in this setting, which seemed to be frozen in time, as we walked in wearing dirty T-shirts and frizzled hair from wearing a bike helmet.

Everyone, of course, looked up as we entered. They were much too polite to stare. They were very friendly and warm and said, "Good morning!" I don't remember much conversation in the room except among us and the farmers (I assumed), who continued with what they were doing—reading the paper, joking with the waitress, or laughing at each other.

When we were done eating, the waitress said, "I will lay your tab down here. Take your time. By the way, the coffee is free."

One of the men at the counter had paid for all our coffee—which one, I haven't a clue. We would soon be crossing the Minnesota state line, and it was a very kind gesture and ceremonial-like farewell to the 900-mile crossing of the Great Plains.

Pulled over early in Williston, North Dakota, to pick up the tent by "general delivery" at the post office. The strategy was to reach Minot, 125 miles away, in one long day. In Minot, the Northern Lights came out in spectacular fashion. Unfortunately, I was too exhausted to keep my eyes open.

6 The Detour: Ely, Minnesota

"Moose burger or caribou dog?"—*Northern Exposure*

I HAD FALLEN ABOUT 3,000 feet in elevation since leaving Browning, Montana. If you had put a marble on the road to follow us, it would have rolled so slowly that it would probably just be leaving the outskirts of Browning by now. I reached Bagley, Minnesota, on July 18. Bagley is 90 miles east of the North Dakota border and is roughly where the forest returns. I had spent the previous night in Grand Forks, North Dakota, crossed the Red River at the Minnesota border in the morning, and spent all day getting back to a forested campground for the first time since leaving Glacier National Park. At one point along the Hi-Line, I had tallied 473 miles without ever seeing a traffic light. I really, *really* wanted to see trees again and wasn't going to stop until I found them.

I had crossed the Laurentian Divide and gone back into the Arctic watershed again somewhere in North Dakota west of Grand Forks, probably around Lakota. The Laurentian Divide dives south out of Canada and makes a long, sweeping arc diagonally across North Dakota, reaches all the way down to South Dakota, and runs back up through the farmland of northwestern Minnesota—as if to gather all the tributaries to the Red River, like a big hand scooping out a chunk of the US's upper Midwest in one last-ditch effort before running back into Canada. Thus, the Red River flows almost due north, which creates some interesting phenomena, especially in the form of chronic spring floods.

First of all, the river hardly has any slope to it. It could just as easily be considered a very wide drainage pathway in a landscape that is relatively flat all around. In Fargo, North Dakota, to the south but upstream of Grand Forks, the elevation is 904 feet. At Grand Forks, 80 miles to the north but downstream, the elevation has dropped a whopping 61 fee—roughly 1 millimeter of fall every 23 feet. By comparison, on my ride across the Great Plains, I had dropped roughly 1 millimeter every

5 feet on average. The ride across the Great Plains was at least four times steeper than the grade of the Red River! Who would've known the Great Plains to be so slanted?

Second, since it is so wide and shallow, the Red River easily freezes in the winter. To make matters worse, as things begin to thaw in South Dakota from the approaching spring, the river to the north still has ice. The ice creates a dam, and things begin to back up. Throw in a warm spell to melt the snow and ice a little faster, along with a rain shower in early spring, and you have a disaster on hand. The river just oozes out of its banks and across a wide expanse of farmland in the Red River Valley—if you can call it a valley—like a spilt glass of water slowly creeping across a table. In the epic flood of 1997, parts of the Red River Valley had water extending 20 miles wide.[21]

As you cross the Red River into Minnesota, you will see combines gathering corn and soybeans from the fields in late summer; Highway 2 is still a four-lane rural thoroughfare divided by a wide, grassy median, which runs so flat and straight all the way to the horizon that the Jolly Green Giant could use it as a bowling alley. But, at Erskine, Minnesota, yet still inside the very edge of the Red River Valley, there sits an oblong-shaped, shallow, pothole-like body of water less than 1 mile across. This body of water is typical of the glacially carved lakes found in the area that were left behind from the previous ice age, of which there are dozens, and is the first indication that we were indeed leaving the Great Plains. It's called Cameron Lake, and it butts right into town at the end of 1st Street, not 400 feet from US Highway 2. I ate lunch at a picnic table across the street from the square, flat-roofed, ascetic post office and its customary flagpole, with kids splashing near the beach. Here I saw naturally landscaped trees for the first time in nearly 1,000 miles.

From there, we caught a tailwind because of cold fronts passing through. A glorious tailwind! It almost felt like you were coasting down a hill. The cold fronts periodically showered rain upon us as they speedily passed by, and then the sun would soon reappear. It seemed as if I had peddled across the entire Great Plains with a slight, dry breeze in my face. Just enough head wind to irritatingly feel like you were pedaling up a slight incline. The shift in winds also signified a dramatic change in venue. Then it was jolting to see spruce trees suddenly appear almost from nowhere as we approached Bagley in the waning moments of daylight. It felt as though we had emerged from a cave that had a fog blowing out of it and, when it all cleared, voilà ... a whole forest of glorious trees!

My main riding partners—a couple named Andre and Elaine—lived in Banff, Alberta. Andre was a carpenter and Elaine was a bilingual customer service representative, and they were on their way to visit friends and family in Montreal. We first camped together just this side of Browning, Montana, on my very first night after leaving Glacier National Park. It was a relief to find new companions so quickly. I had lost them when I stopped in Culbertson but caught up with them again in North Dakota. It was uncanny how we knew where each other would be stopping for the night. I guess we all had the same goals and intuitively found the same place: somewhere that is safe, clean, and cheap, and has water. A tree or two would be icing on the cake. I always pedaled alone, but we often stopped for breakfast together if we could find a café, and we found each other again at night—like birds in flock that come to roost just before the sun goes down.

Cafés were few and far between in the plains, but now the population density was starting to increase—yet another indication of changing conditions. Between Grand Forks and Bagley, we had eaten at a café in Fisher, Minnesota, a tiny town with 435 people that sits on a bend of the Red Lake River, which is another wide, flatland tributary to the Red River.

After leaving Bagley, I got the $1.99 special at a café along the highway 1 mile east of the tiny town of Wilton, population 204. Wilton is no bigger than most towns I had gone through in the Great Plains, but the overall density was increasing as evident by the sprawl of an increasing number of tiny communities scattered along US Highway 2. Paradoxically, in the Great Plains, it was as though the population were at such an irreducible minimum that they could not afford any sprawl, and the towns there were more concentrated. Nelson County, North Dakota (Lakota), the one to the west of Grand Forks County and where we ate breakfast in the Chevy dealership, has a population of 2,968 people. Clearwater County, Minnesota (Bagley) is almost identical in land size and has a population that is three times larger, yet Lakota is only half the size of Bagley. In the Great Plains, you had to fill several water bottles at each town in order to traverse the long, hot gaps between each one; here, you almost didn't need one. In the Great Plains, you had to plan way ahead, study the maps, and make sure everything was tuned on the bike, and you still never felt at ease navigating the long distances.

For me, the preparation and vigilance were intense. I often pedaled all day, fell in the tent with barely any supper, and was soon sound asleep. Up again at sunrise and repeat. But, in Bena, Minnesota, population 116, I was able to play the jukebox and drink a beer at lunch for the first time since leaving Glacier National Park.

The Mississippi Headwaters State Forest surrounds the stretch of US Highway 2 between Bagley and Bemidji. Somewhere along that stretch of highway, where it had seemed to close in and narrow down to a two-lane road bordered by quaking aspen trees, I had finished skirting through a piece of the Arctic watershed for a second time as I left the Red River Basin. Although the true beginning of the Mississippi River begins as a trickle in Itasca State Park, it begins to pick up enough steam to float a canoe somewhere just south of US Highway 2. The river actually meanders *north* out of Lake Itasca, and then begins a broad, slow, patient, 30-mile-long, sweeping curve to the east before finally turning south while connecting several natural glacial lakes, like pearls on a necklace. There are more lakes than there is river.

Kind of like, "Do you want a little spaghetti with that sauce?"

In Bemidji, Minnesota, at the exact spot where the Mississippi River connects two of those clear, turquoise lakes, Paul Bunyan Drive squeezes between them and bridges a piece of the string that is less than three city blocks long and so narrow you could long jump across it. The bridge is so indistinct that, if you stand at the edge of the water underneath, you must duck so that you don't hit your head. If I wasn't sure I had made it through the Great Plains, I knew now because there is an 18-foot-tall statue of the street's namesake and Babe the Blue Ox in the park around Lake Bemidji, the larger of the two lakes that lies on the north side of the bridge. Elaine snapped a picture of me leaning against Babe with my arm stretched out and my hand resting just above her elbow.

The water below the bridge is still flowing north, but then the river abruptly dumps out of the east side of Lake Bemidji, like the spout on a teapot, while the lake itself continues to expand to the north nearly to the horizon. I had a broken bar on my back rack that needed to be welded, and Bemidji was also the first place in two weeks whereby I left US Highway 2 to go into town for errands but reentered the highway farther east, like when a race car drives through pit row.

Very soon, I would be leaving my riding partners for good. The magic of Minnesota has haunted me my whole life. And the town of Ely, Minnesota, had become my ghost house filled with Casper-like spirits. Ely draws me in like the Sirens of Homer's *Odyssey*. I had always worked in Ely any time I had a break from college up until the moment I pedaled off the Pacific shoreline of Washington just four weeks earlier. Over the past spring, I had felt a twinge of betrayal as I planned this quest that I was now on. Ely still teases me and, when the seasons begin to change, I wonder, "Are the lakes thawed by now? Or are they thick enough to ski on yet?"

As my friends were planning to continue east, I would detour through the heart of this magnificent land dominated by meandering waterways, bogs, and balsam fir. It is where my second career in environmental science took root. It is where I would later return in my late 30s to gain the education I needed to begin that second career.

The day after camping in a birch grove at Deer River, Minnesota, Andre, Elaine, and I ate breakfast in the town of Cohasset, where the Mississippi River begins to toy with the idea of turning south and recrosses US Highway 2 through marshland filled with black spruce and speckled alder, and where tamarack trees grow in the upland edges of the wide swamps surrounding the wild river corridor, which is as narrow as the nearby streams that feed it.

I broke tradition of riding at my own pace and stayed alongside my companions during that last day, together for one final, drawn-out farewell. After eating lunch in Grand Rapids—the spot at which the US Army Corps of Engineers put their first dam on the Mississippi and turned it from a frontier waterway into one that was to be tamed by man—I turned north and left US Highway 2 for much longer than just running errands. And, for the second time, I waved goodbye to closely bonded friends whom I have never seen again.

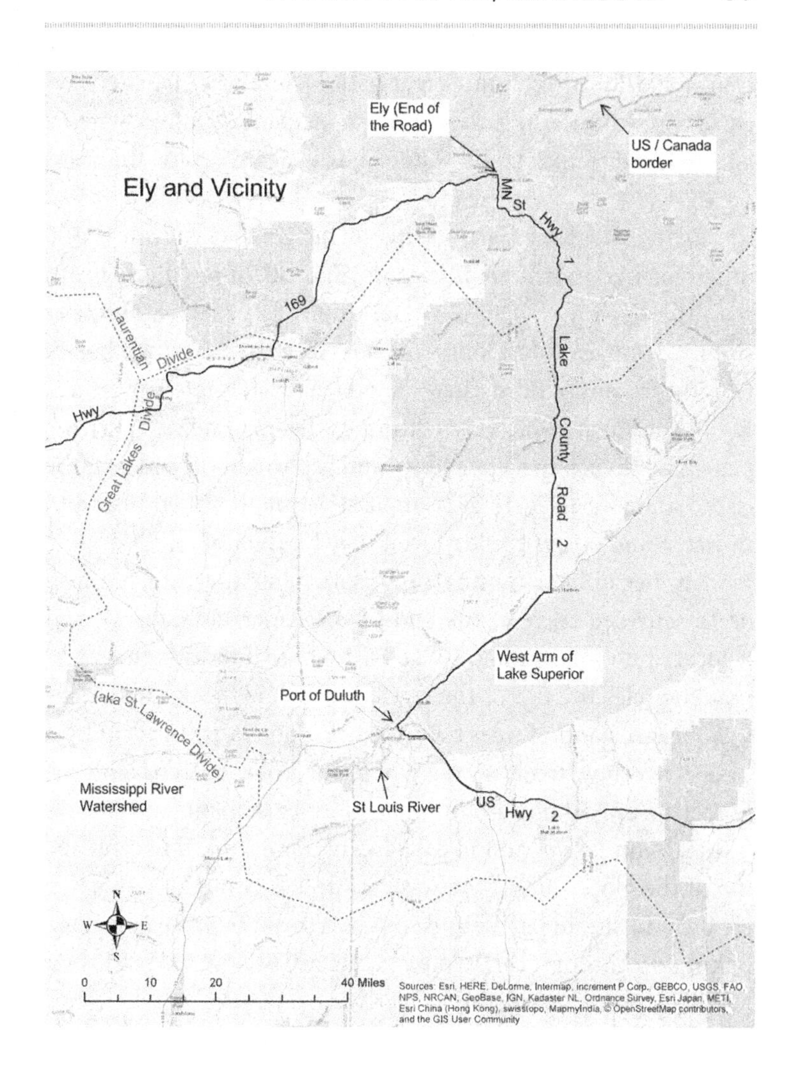

ELY, MINNESOTA, IS ANOTHER one of the few places in the continental United States where the water flows neither to the Pacific Ocean, the Mississippi River, the Atlantic Ocean, nor the Gulf of Mexico. Somewhere along Highway 169, just north of the town of Virginia, Minnesota, I climbed 2 miles to get over a 400-foot-high hump where the Laurentian Divide stays close to the border but wiggles back inside the United States. I had fallen into the Arctic watershed … again … for the third time.

Ely is called the Far Northland for good reason. It is 300 miles farther north than Toronto and 30 miles more north than the tip-top of Maine. For example, as part of a publicity stunt for tourism development, one year the Ely Chamber of Commerce announced that Canada had attempted to annex Ely in the April Fool's Day edition of the local newspaper. The paper did their journalistic duty and reported both sides of the issue; some locals were as indignant as cats that had been put in water, and others couldn't wait to leave the union.

Ely began as a mining town to exploit the valuable iron ore discovered there at the end of the American Civil War. It is located on the southern edge of the boreal forest—the forest that encircles the top of the globe just below the Arctic Circle. It is the last stand of trees before Earth's landscape turns to tundra—a scrubby, treeless area where the polar bears roam. The boreal forest canopy is dominated by balsam fir and spruce trees. If you could ride in a spaceship and look down on the top of the globe during the wintertime, you would see this forest around the upper latitudes as a narrow band of greenery, poking up through the whiteout like a Christmas wreath that had been placed on the top of Earth.

Thus, Ely morphed into a logging town to harvest this other valuable natural resource after the mines had become unprofitable due to depletion of the richest veins and from global competition. Although logging still occurs today—mostly through managed plantations by the US Forest Service—Ely primarily has now settled into being the gateway to the Canoe Capital of the World. There are 2 million acres of roadless wilderness with thousands of pristine lakes, portage trails, wetlands, and streams all meandering northward towards Lake Winnipeg, the "sixth Great Lake," the same destination as the Red River.

This huge expanse of managed wilderness incorporates the invisible border between the United States and Canada and is equally divided on either side. In this part of the North American continent, the border between the two countries follows the traditional navigable waterway, which was the most direct route into the interior of the continent from Lake Superior and most often used by the French-Canadian paddlers who traded for furs with the Algonquin indigenous people. It was an amicable relationship—so much so that many of the traders married into the tribes; some may argue that they only did it to gain favor with the tribes and push out competing traders.

Nonetheless, these 18th century European entrepreneurs had no interest to settle the land and extract it from the native people. And that kept the peace—if not between competing Europeans, then at least between the two cultures. If you paddle into this public area today, it is as remote and pristine as the day the first voyageur out of Montreal entered this treasure trove of natural resources. In fact, some Ojibwa tribes still use the area for sustenance hunting and fishing. The water is as clean as any on the planet.

It always seemed silly to me that border regulations even existed, but I guess every bureaucracy has to protect its turf. The better solution, it seems to me, would be to draw a circle around the 2 million acres in both countries and have controlled access by an international police force that consisted of US agents on the US side and Canadian agents on the Canadian side. Instead of the point of regulation occurring at the political line that splits the wilderness in half, the point of regulation would occur at the perimeter. Both countries would be in constant communication, and everyone wins: increased efficiency for both governments, less hassle to the wilderness visitor, and increased security for both countries.

There are only three national, high-adventure programs run by the Boys Scouts of America's home office in Irving, Texas, and one is located in Ely. In the summer, I was a "Charlie Guide" (there are women guides too; the term is named after an early founder from 1923) at the Northern Tier High Adventure Base and took groups of eight on backcountry fishing and canoeing adventures. In the winter, we ran a program called "Okpik"[22] from the day after Christmas through February, where kids would pull sleds with cross-country skis or ride behind sled dogs, and camp out in polar domes made from hollowed-out piles of snow.

In January, you can ski across the same lake that you paddled across in June, and the director liked to say, "You know, the temperature in Celsius and the temperature in Fahrenheit is the same at minus 40."

Diving into one of those crisp lakes—whose surface is reflective enough to recognize your own face blemishes on a still, hot July day—is as much relief as a cool breeze on a cloudless day in the Arizona desert. Yet, the lakes of the Canoe Capital of the World became perfectly flat, white football fields after Thanksgiving. In fact, if you want to traverse from Ely into Ontario, the *only* way is by canoe in summer or on skis in winter.

During the one week I was off the bike and volunteering at base camp—where we had electricity, a dining hall, cabins, and a shower house, and where crews were outfitted before going on the water / ice—I took one day to get out to the North Hegman Lake pictographs. Northern Tier is located on Moose Lake and, being only 6 miles via continuous waterway to the Canadian ranger station at Prairie Portage, it just wasn't that convenient to get over to the Hegman Lake area. I had never seen probably the most visited and arguably best set of pictographs in the Boundary Waters / Quetico Wilderness. It is a free, outdoor art gallery—no membership required. Fortunately, they are accessible via a dirt road out of Ely, two short portages, and two short paddles. The pictographs were drawn with red paint, most likely made from iron ore, among other things, out of the very rocks they were painted on by an Ojibwa artist many centuries ago. The paintings are on a flat rock the size of an interstate billboard and overhang the shoreline in one corner of the lake.

Here is my attempt at an interpretation: The paintings depict a human with long, outstretched arms and hands the size of basketballs, possibly symbolizing a spiritual character; a dog or wolf with a very long tail, possibly a caricature of the animal; and a detail of a bull moose without any exaggerations and with an impressive rack of antlers. The spirit character is not stick-like, but it is not detailed either, there being no discernable eyes, hair, or ears. The body tapers uniformly, almost wedge-like, from the chest to the legs. Below these three images is a horizontal line with a slight dip that may symbolize the ground upon which the two animals are standing, with the spirit character being slightly above them. The animals are correctly proportioned to each other while the spirit character is nearly double the size of the moose. To the right of the head of the spirit character appears to be a small school of fish. And above the spirit character are three rudimentary canoes: a single, curved lined representing each one, and two of them with very small, blocky, stick-like figures.

After the week of visiting old friends, I was ready to move on through my 200-mile detour. I rode due east early one morning on the only other paved exit out of Ely: MN State Highway 1. After a while, I turned right onto Lake County

Road 2 and rode it through Superior National Forest, all the way to the shore of Lake Superior, where I entered the small town of Two Harbors—the only sign of civilization since leaving Ely. I made another right turn at Two Harbors and rode the shoreline all the way to the port of Duluth. On my way to Duluth that day, I finally left the Arctic watershed for good because, somewhere along the road to Two Harbors, I had crossed into my fourth major continental watershed and gone over the Great Lakes Divide.

Lake Superior is considered an inland freshwater sea. It drains east through all the other Great Lakes, and its water eventually reaches the Atlantic Ocean via the St. Lawrence River. It takes an average drop of water 321 years to do that.[23] Lake Superior could hold the water of all the other Great Lakes combined and contains 10 percent of the world's surface freshwater that isn't frozen. It has a surface area of 31,700 square miles, an average depth of 483 feet, and can have hurricane-like waves on a treacherous day. Gordon Lightfoot makes this situation vividly clear in his tragic song, *The Wreck of the Edmund Fitzgerald.*

Lake Superior is so enormous that it becomes indistinguishable where the sky ends and the lake begins; they appear to blend together as one. The rural road into Two Harbors sits up high above the lake on a plateau and, on a clear day, you feel as though—if not for the white birch and quaking aspen trees—you might be flying in the sky because the blue surface of the lake sits just below your line of site and the sky just above.

Large bodies of water such as Lake Superior are heat capacitors; they store heat in the summer and release heat in the winter; therefore, they moderate the local climate. Ely is far enough inland from the lake that it can be as much as 10 degrees colder in the winter and 10 degrees hotter in the summer as compared to Two Harbors or Duluth. When I left Ely, I was in a short-sleeve shirt on this clear, summer day. Even before the lake became visible, I could feel the temperature dropping; by the time I was within 3 miles of the shore, I had stopped already to put on a fleece.

IN MID-2021, GREAT RIVER ENERGY—a Minnesota electric cooperative similar to MEAG—was trying to close down a coal plant, not because of the climate crisis but because renewables had become a cheaper way to provide power to their members. Keeping the plant open was costing them about 8 cents a kWh. To put that in perspective, I am paying *retail* 9 cents a kWh (plus some franchise fees) from MEAG. North Dakota, not wanting to lose the coal-mining jobs that were supplying the plant, started playing hardball and put a moratorium on wind power, so that the transmission line to Minnesota could only carry coal-fired power. To solve this problem, Great River Energy sold the (almost) worthless coal plant for $1 and the power line for $225 million—I kid you not.[24]

This situation is akin to trying to hold back the inevitable tide because a renewable energy grid is happening as we speak, and you don't have to burn things to get the energy we need to run our economy. Digging fossil out the ground and setting it on fire is an unsustainable proposition. The good thing now is that the denialism on climate science has been pushed to the fringes.

The first retreat was by the fossil industry, which decided to take the approach that the climate actually is changing (rapidly for Earth time), but we aren't causing it. Then, when that position was overrun, they decided that it isn't that big of a deal. This position has been overrun too, so the next statement was that we can't transition away from fossil too fast because it would hurt the poor and raise energy prices too high. Ten years ago, they even said that renewables like wind and solar would never be able to compete with coal.

When my family put a 6-kilowatt (kW)[25] solar system on our roof in 2018, the cost of a battery to store the excess could not be justified. We got tired of sending the excess to the grid for about a penny per kWh; three years later, battery prices had fallen so low that the addition of a 10-kWh battery was a no-brainer. To put that in perspective, our Chevy Bolt has a 60-kWh battery with 220 miles of range. The house battery is just big enough to act as a reservoir for when a cloud rolls past or house demand increases beyond solar-panel capacity and to take up the excess when we are overproducing. Not having a battery along with solar panels is like trying to supply a community with water from a running stream with no dam. It acts as a "buffer." If the power goes out, the battery allows for lights and computers to run without blinking. But it isn't there to run the energy hogs like heating water (showers) or drying clothes or running the A/C or charging the car. (If I had to rank those hogs, the first two are tied—30-amp breakers—and the second two are close—20-amp breakers.) But we now have less invested in our solar/battery system ($21,000) than we do our car ($24,000).

I am not saying that technology is the only answer. It helps, but we need more than that. There is no magic wand, and I think it will take hard work by all of us to solve what is essentially a humanitarian crisis more than an environmental one. We should probably be investing like crazy into things like transportation *diversity*. You wouldn't put your entire retirement portfolio in one stock, would you?

The one low-tech thing in which I have invested, which has given me the biggest bang for my buck—or, as a Wall Street analyst might say, a return on investment—is the spray foam insulation in the attic and upstairs outer walls. When we added the second floor, the contractor said that spray foam ($3,000) would cost twice as much as fiberglass insulation ($1,500).

I asked, "What difference does $1,500 make inside an $80,000 project anyway?"

Besides, it allowed us to cut the size of our heating and cooling system in half, and we were able to install the smallest heat pump (1.5 ton) that could be found on the market.

So, the latest position that the fossil industry has retreated to is that we should wait until we have the technology to capture carbon out of the air and sequester it somewhere else before we abandon our stranded assets sitting in the ground. Now *that* would be waiting on the magic wand, and I don't think we have any more time left.[26]

YEARS LATER, WHEN I was living in Ely, I realized that you couldn't get through the winter without consuming high-density food and lots of it. Even though you could ski on a sunny day on top of frozen lakes in a T-shirt and sled dogs would overheat when it got above 30 degrees F, nights could be brutal. Either way, you were burning through calories.

I decided that I would buy a rifle and acquire a deer-hunting license well before the fall hunting season. In Ely, even on a sweltering 80-degree day in July, you were always planning for winter.

I would ask myself, "Do I have enough birch stockpiled for the wood-burning stove?"

That meant, "Do I have enough stockpiled *12 months in advance*?"

The split wood that was dumped in your driveway or front yard was cut from a live stock of birch or green ash, the most calorie-dense wood that burned the hottest. But only if it was dry. Thus, the 12-month "seasoning" process for the wood was required in order to keep a house warm when it was 25 degrees below zero many nights; otherwise, the heat from the stove was wasted on dehydrating wood and not heating the house sufficiently. In the summertime, you might be thinking you had enough wood for *this* winter, but you were also thinking about the winter beyond.

I started wondering if I had enough meat to get me through the upcoming winter. I drove to the town of Virginia and found a nice, used Winchester .30-30 for $200. It took me about 20 minutes to decide.

"I'll take it!"

I had to submit a background check, and three days later I was back in Virginia to pick up the rifle.

"Need a scope?" the clerk asked.

"No, sir, but I'll take a 20 box of ammunition."

I then went to the hardware store in Ely and figured out the nuances of the current year's hunting regulations, which are always based on what zone you wanted to hunt and on what the deer-mortality surveys found in the previous spring.

I had worked one of the deer-mortality transects for the Minnesota Department of Natural Resources. You basically have a 1-mile compass bearing where you walk through a wooded area on public lands and record any deer carcasses you find and how the site looked: Were they killed by wolves or did they starve? If they weren't killed by wolves (which most aren't because a wolf chase is only successful in about one out of every seven attempts), then they usually died from starvation in early spring.

By April, if a deer hadn't quite put enough fat stores away in the previous leaf season, he is out of gas. When your car is out of gas, you pull over and walk to a gas station to fill a gas can, so you can start it again. When a deer is out of gas, it dies. A deer with depleted fat stores in the bone marrow would have died from starvation. Sometimes these bones are left after a wolf kill and can be examined. Fat stores usually mean wolf kill; no fat stores usually mean starvation. If most of the deer died by starvation, then the hunting quota is reduced. But if most deer are killed by a wolf pack, the hunting quota remains about the same.

The reason for the difference to changes in the hunting quota is because long, hard winters have a much larger effect on deer populations than wolf kills. Deer populations that are hammered by hard winters fall dramatically and then recover over a few years' time. Wolf kills are a near constant. And wolf populations are dependent on deer populations, not the other way around. In other words, the local wolf population would go extinct long before the deer population fell to any threatening level. It takes about five times more deer than wolves to sustain one wolf family. If the deer population falls below the five-times multiplier, the wolf population reaches a tipping point at which it cannot sustain itself and would eventually decline to zero. It's a basic ecological principle that energy flows from the bottom up and not the other way around.

I bought a license in a nearby hunting zone that allowed the taking of both does and bucks with a rifle. Besides, I had a friend who owned a home in the same zone and was at his house on the first day of hunting season. Fortunately, my friend was in South America, visiting his girlfriend, and he needed a house sitter. His house happened to be on a wooded tract of land and well outside the town limits.

I got up before dawn and made a pot of coffee. It was already below freezing in the early twilight of late November. After having my coffee and watching the sun slowly creep above the horizon, I sat in a chair on the gravel driveway, which faced a wooded field on a power line about 75 yards away.

I thought, "That has to be a good place for deer to wander into, and it also gives me a good line of site."

I sat for about an hour and then got impatient. I walked back inside to get another cup of coffee. When I came back to the door, there was a big doe standing on the power line. I leaned against the door frame, slowly put the gun up to my shoulder, looked down the barrel, sited on the deer's chest … and pulled the trigger. It bolted!

"Did I miss?"

I ran to the power line where it had been standing and carried the rifle with me. It was loaded with two more bullets. At the edge of the woods next to the field, I saw bright red blood on a birch sapling. I followed it to another sapling with a blood splotch and continued for another 25 yards. There it lay.

I let out a sigh of relief and thought, "I don't have to shoot it again."

It was dead. I had never fired the gun before and have never fired it since, not even to site test it on a firing range. And I swear on the Bible that the box of ammunition still sits in my underwear drawer with 19 bullets in it.

DURING THAT WINTER IN ELY, I had 60 pounds of boneless meat in my freezer. I had hamburger, steaks, and roasts. At Christmas, I made jerky and gave some away as gifts. But I thought it was ridiculous to have a freezer running in the house when it never got above freezing outside for at least three months during the year.

I eventually decided to pack everything in a wood box and set it in a snowbank by the side door to the driveway. I wasn't worried about people stealing my winter stores. But the perennial wintertime resident of Ely—the raven—was probably the only thing that would have enough intelligence to find a way to open the lid and cache it away for himself, and the only one with enough curiosity to want to know what's inside. I secured the lid to the box and turned off the freezer.

I estimated the doe to be about 200 pounds on the hoof. In Florida, that would be a huge deer; for the cold climate of northern Minnesota, it is only average. The farther north you go in the northern hemisphere, the larger the deer get. It is another ecological principle that has to do with heat conservation: The larger you are, the less surface area (skin) there is to lose heat in proportions to your overall body mass. For example, the inside of a golf ball could hold about 41 milliliters of water and has 11 square inches of covering. A basketball could hold 180 times more water than a golf ball but only needs about 25 times more covering than a golf ball does. If you were to heat the water before you put it in each ball, the water inside would eventually cool down to room temperature; the water in the golf ball would reach room temperature significantly quicker though.

You might ask, "Well, how in the world do tiny chickadees, which don't migrate, survive a boreal winter?"

They have other wintertime adaptations, such as the ability to go into hibernation mode at night, whereby metabolism slows down and energy is conserved. But what's interesting is that the black-capped chickadee, the northern version, is about 10 percent bigger than a Carolina chickadee, the southern version. Elevation is analogous to moving north, climatically speaking, and biologists have determined what elevation in the Great Smoky Mountains National Park separates the two chickadee species.[27] This delineating elevation line was roughly at 4,000 feet in 1952. I could imagine this delineating line migrating upwards until the black-capped chickadee has been squeezed out of the Smokey Mountains.

IF YOU GO TO the International Wolf Center in Ely, Minnesota, you will see a stuffed wolf on display that has been commemorated: Wolf 919, a 75-pound male that lived to be at least six years old but no more than 10 years old. No matter how old he lived, he got to pass on his genes to the next generation, and he certainly beat the odds. The average life span of a wolf in the wild is four years. He was scrappy and a survivor. I know because I tracked him during the winter months of early 2008.

One day, during the third week of March, I placed the telemetry device on the Wolf Center Director's desk and said, "Found him."

"You did?"

"Without a doubt. Clear, strong signal. The hard part was getting a triangulation."

"Why? Where were you?"

"All the way to Twin Lakes. Unplowed!"

We had gotten a 6-inch dump the night before. The major thoroughfares are usually plowed immediately, sometimes even in the middle of the night. When exactly they get plowed depends on the amount of snowfall, the forecast, and available resources from the state. Some local roads are plowed by the municipality, or sometimes a local resident will have a plow on the front of his or her truck. The Twin Lakes Road has few residents, and sometimes the road may go a week without being plowed.

Wolf 919 had been collared by the United States Geological Survey (USGS) when he was a juvenile with his home pack. But the USGS had run out of funding, and he had not been tracked regularly on the ground in three years. Every once in a while, the USGS would send up a plane to log a position on the wolves that had been collared or check for a mortality signal; if the collar doesn't move for 24 hours, it will automatically send a steady ping, which is distinctly different than the normal transmission that comes from a live wolf. Global Positioning System collars were still too expensive, and the cost could not be justified at the time Wolf 919 had been trapped. In the meantime, it was known that he had left his home pack, found a female, and was well on his way to forming a new pack.

The Wolf Center in conjunction with Vermilion College, the local community college, had taken over duties in order to determine the boundaries of his territory. We had gotten about a dozen locations on him at different times of the day, but we needed at least 30 locations to be confident in drawing an accurate polygon. We had a rough idea of what the borders might be, due to the information we had on nearby packs; wolves from one pack do not let wolves from another pack come into their territory. And we had information according to other constraints such as roads, which often create pack borders. But we had never expected him to be as far west as Twin Lakes. It had been almost two weeks since we last got a reading on him. I had gone down the Twin Lakes Road almost in desperation.

"Yeah," I said to the Wolf Center director. "I had to drive straight in, all the way to the lake, without turning the wheels too much."

The director looked puzzled. I owned a manual-transmission, front-wheel drive Mazda Protegé. It was nearly foolish to do what I had done. I figured that I would be walking the two miles back to the main highway.

"I got the first signal standing on the lake. He was really close. Must have been on the lake somewhere."

Same lake along the Minnesota-Ontario border

January

June

The director just looked at me and nodded.

If you are not familiar with Ely, Minnesota, the lakes stay frozen—thick enough on which to drive a tractor trailer until May. In fact, the town actually plows a "road"—with light posts and all—across Shagawa Lake every winter. Everyone, including school buses, use it to cut the corner when going from the northern "suburbs" into town. The word "suburb" is used lightly because Ely is smack-dab in the middle of the wilderness. Ely is so remote that the "End of the Road Radio" (WELY AM 1450 / FM 94.5) is one of the few radio stations left in the United States that are mandated by the Federal Communications Commission to provide official emergency information to local residents in times of crises.

"Then I put it in reverse," I continued, "and drove backwards to where West Boundary Road forks west. I stopped the car, put on the emergency brake, and walked down West Boundary about 50 yards to get the second signal. Not the best triangulation, but it worked."

When you use telemetry, one signal cannot pinpoint where your subject is. He is obviously somewhere on the line of the signal, but where on the line? So you have to move away, preferably as perpendicular to the line of the first signal as possible. You record the compass direction of the two signals and draw them on a map, and where they cross is your location point.

I got back in the car after getting the second signal, held the driver door open, slowly released the clutch, held my head out the door looking backwards, and steered the car in reverse. I figured I would go as far as I could, and then start walking. I could hardly believe it when I finally saw the dark, plowed pavement of the main highway. I had made it!

There is a cartoon posted on my office wall that shows two guys looking out an office window at a massive ice sheet.

The caption says, "A looming glacier certainly puts things into perspective."

One day, the director of the Wolf Center told me, "You know, John, politics always trumps science."

This statement coming from a scientist who had a small part of the decades-long research that eventually changed people's view of how predators interact with the ecosystem. The work changed public opinion so much that the wolf has been brought back from the brink of extinction in the Upper Midwest. The biggest threat to the wolf in Ely now is being hit by a car.

WHILE I WAS IN ELY and working at the International Wolf Center, I had also been working and taking classes in ecology at Vermilion College. The two institutions collaborated and shared resources, and one of my jobs was to paddle out and retrieve the telemetry collar from wolves that had died. We would get general coordinates from the plane as to its general location. According to how far they were in the wilderness, sometimes we could paddle out, hopefully find it along the closest portage trail with the telemetry equipment, and get back home in a day. At other times, it would be so far out that it was impossible to paddle there, retrieve it, and get back home before dark, so we had to camp overnight.

Once you got within 50 meters of the deceased wolf, the telemetry would ping clearly in almost any direction, and you couldn't tell which way to walk. And the dead wolf might be hidden among the thick underbrush. Usually, when you got really close though, turkey vultures would fly up and reveal the secret spot.

Another one of my projects was to restart the Superior National Forest's Boreal Owl study. The biologist who had designed the study, and who five years earlier had put up 100 nesting boxes on the trunks of white spruce or balsam fir trees about 20 feet above the ground, had not been able to monitor them in the past two years. I was given the coordinates of all the boxes and was directed to check each one during the nesting season in the first week of May and then write a report.

Boreal owls don't build their own nests and therefore utilize natural holes in trees or ones that had been carved by woodpeckers. Nesting boxes were a good substitute—kind of like putting up a bluebird box—and you could place them randomly throughout the forest. The boxes weren't that big because the boreal owl is no bigger than a robin. The boreal owl was a species of concern due to possibly declining populations, so we wanted to see if we could figure out what kind of habitat in which they preferred to nest. It helps inform managers of the Superior National Forest how to balance the competing demands between commercial logging permits and forest conservation.

The problem is that, in the three years the boreal owl had been monitored, a similar species that seems to be doing okay—the northern saw-whet owl—had always been found in any occupied box, and not one boreal owl had ever been seen using them. And there were never many boxes that were occupied in the first place. Therefore, data were severely lacking to inform any conclusion. When I checked them, seven of the boxes had been severely damaged or were lying on the ground; of the 93 left, five had been used.

One day, I called Dave, the national forest biologist, and said, "There is a boreal owl in number 73."

"Really? Are you sure?"

I replied, "Yeah. It has the right markings. Please go verify."

A day later, he called back and said, "Yep … it's definitely a boreal."

About two days after that, I called again. "There is a boreal owl in number 12."

He hadn't known me very well and still wasn't 100 percent confident in my ability to distinguish it from the northern saw-whet owl.

"Please go verify," I requested.

The next day he affirmed, "Wow! It's a boreal."

7 The Great Lakes Watershed

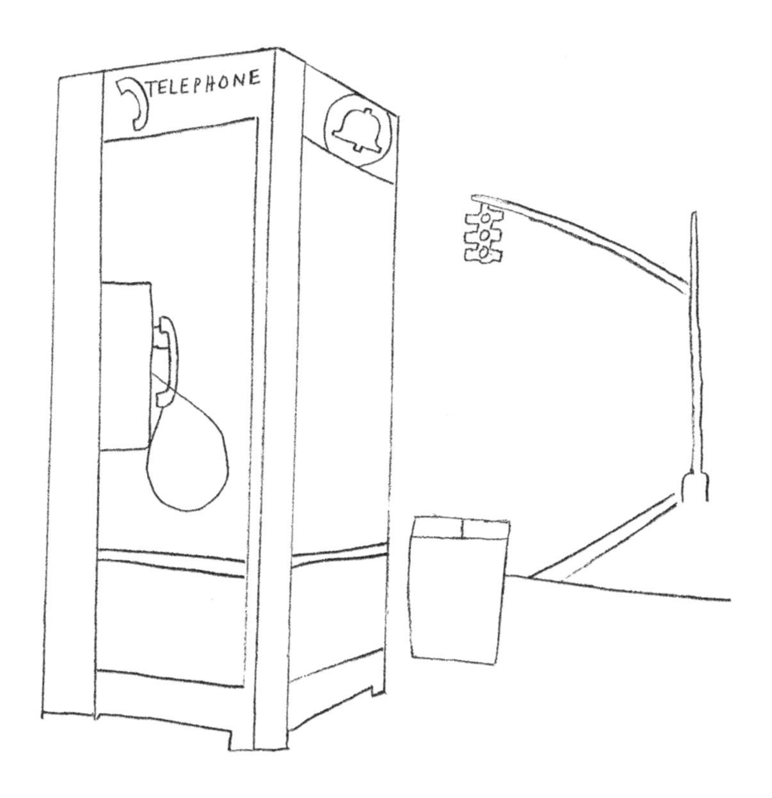

"Toto, I've a feeling we're not in Kansas anymore."
—Dorothy, *The Wizard of Oz*

LAKE SUPERIOR'S ELEVATION at full pool is 600 feet above sea level. It drains into Lake Huron, elevation 577 feet, via the St. Marys River. Lake Huron in turn drains into Lake Erie, elevation 569, via the Detroit River. It then drains into Lake Ontario, elevation 243, via Niagara Falls. The interesting thing is that Ely is only 48 miles from the shore of Lake Superior as the crow flies, but Lake Superior does not capture any of its water. In some areas around Lake Michigan, the area of drainage from the shoreline only extends outward across the land for a mere 9 miles.

Along the shoreline that I rode to Duluth, majestic cliffs 100 feet above the water hold historic lighthouses but contain almost no surface tributaries flowing towards the lake. Yet, these lakes hold 20 percent of all the surface freshwater on the entire planet, and 500,000 gallons of water flows from Lake Superior into Lake Huron every second.[28] The surface area of the five lakes is a combined 94,000 square miles.

The region is so large that, although I would be inside it long enough to see only three of the five lakes, it would take me two weeks to wind through the domain of the Great Lakes watershed. I stayed over in Duluth for two nights and then crossed the bridge over the St. Louis River Delta feeding Lake Superior (also known as "Gichi-gami" or "great sea" by the Ojibwa) and into Wisconsin on the last day of July. I cut right through the heart of the Great Lakes region on the large peninsula jutting east, which hangs off the mainland of Wisconsin (the Upper Peninsula, known as the "UP"). Even though it lies within the political boundaries of Michigan, its only connection to mainland Michigan—the Lower Peninsula—is over a 5-mile-long, four-lane-wide, man-made bridge. And, although the Great Lakes region covers such an enormous area, in the UP, I could easily bump from the shore of Lake Superior to the shore of Lake Michigan in a single day.

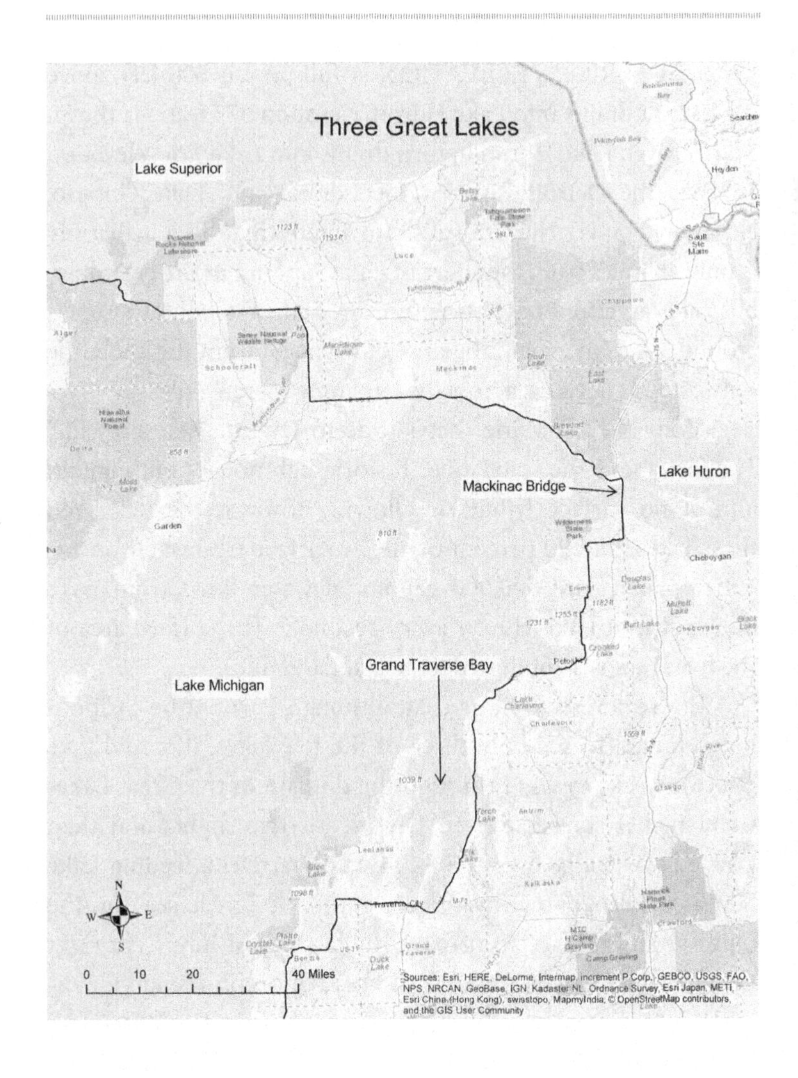

Thirty-seven days and 1,707 miles after getting the handwritten note from Biff in Glacier National Park, I finally reached Traverse City, Michigan. Two days before arriving in Traverse City, I had camped along the *south* shoreline of the UP, near the beaches of Lake Michigan at Hog Island State Forest Campground. And the day before that, I had camped along the *north* shoreline of the UP, overlooking Lake Superior at Bay Furnace Campground inside the city limits of Christmas, Michigan. The town of Christmas, population 400, owed its existence to the fact that they once had a huge stone-blast furnace where ships plying the Great Lakes and carrying iron ore from the mines near Ely would stop and convert their raw loads into pig iron—an intermediary, mostly unusable, iron product but lighter to ship.

Riding between the two Great Lakes campgrounds, I traversed the breadth of the UP during an 84-mile day. On the crossover from Lake Superior to Lake Michigan, I had stopped for lunch in Blaney Park—a tiny town with three antique shops, about a dozen houses, and a bed and breakfast, surrounded by white spruce trees. The rear wheel was starting to stress from the weight of my gear on the back rack, and I had spent most of the day riding with one broken spoke. The front tire doesn't wear quite as fast as the rear one; so, during the last hour of sunlight on that crossover day in the campground along the wooded shoreline of Lake Michigan, I rotated the tires while the rear wheel was off to replace the broken spoke.

The next day, I made the big turn south and left US Highway 2 for good at the Mackinac Bridge. Out in the distance, I noticed the nearly 5-mile-long suspension bridge—known as Big Mac—at least 45 minutes before actually arriving at the Straits of Mackinac[29] because I had a clear view over the broad, open expanse of Lake Michigan while riding due east. The bridge itself ran perpendicular to my path of travel and just to the right of my line of site, as it connects the two peninsulas of Michigan. The peninsulas had been connected by ferries until the bridge was opened in 1957. Although its main span between the two towers is 400 feet shorter than the Golden Gate, its total length is nearly three times longer. When

I reached the bridge—a busy four-lane thoroughfare carrying I-75 traffic—the Michigan Department of Transportation (DOT) would not let me cross it on bicycle (not that I would want to). In fact, Mackinac Bridge Authority has a 24/7 "Drivers Assistance Program" that even provides drivers for *motorists* who are uncomfortable with driving across the Mackinac Bridge. I paid them $1 to put my bicycle in a DOT pickup truck and carry me across.

In Traverse City—just as I had in Culbertson, Montana—I stopped at a pay phone booth. This time, the price of a phone call had more than doubled, so I put in a quarter before dialing the number. It seemed awkward, and I didn't know what to expect. But Hanz had been expecting my call and almost immediately gave me directions to his house as though, like Biff, he had known me my whole life.

Hanz lived at the bottom of Lake Michigan's Grand Traverse Bay along a flat, semicircular road that hugged the shoreline. A large, prehistoric ice sheet was the architect of this geological design, which created the slight spread to the 30-mile-long "pinkie" finger of the "hand" of Michigan's Lower Peninsula. The ice sheet has come and gone many times, and it actually sat on top of the pinkie just the same, but the bedrock there was much harder than the surrounding earth and the ice couldn't quite grind it down enough to make it disappear beneath the waves when it last retreated.

The bay is quite impressive and, as you stare out across the water, it seems as though you are looking at one of those seas described in the Old Testament. When I arrived, the wind was blowing in from the north under a sunny sky, and small white caps were lapping up to the shore. Houses had been built on the opposite side of the road away from the shore, thereby leaving the bay side of the road looking like a preserved lagoon. It was a public space with such clear water that you could see the sandy bottom of Lake Michigan. Although no houses had been permitted to be built along the water, homeowners were given a permanent easement, if they wanted, to build a dock in the shallow water of the bay, yet anyone could swim there. It was a fantastic public-private partnership.

When I walked into Hanz's house, he immediately showed me where the shower was. He then took my dirty clothes and put them in the washing machine. After cleaning up, he drove me over to an orchard to pick apples.

"Sorry, but you are too late for the cherries," he said.

Traverse City is sometimes called the Cherry Capital of the World. Five counties surrounding the bay, covering an area of 2,981 square miles with a population of 170,000, make up 40 percent of the tart cherry production in the United States.[30] Like Lake Superior, Lake Michigan is a large enough body of water to moderate the local climate. It soaks up heat in the summer to keep the cherry trees cool and protects them from intense cold in the winter. This climate and sandy, well-drained soil make it a primo area for growing cherries. It is the major driver of the economy and dominates the culture of the region.

Very near downtown Traverse City, at the bottom of this bowl-shaped shoreline that forms the southern boundary of Grand Traverse Bay, a narrow 15-mile-long, 1-1/2-mile-wide peninsula juts upward into this vast bay from the mainland. It was also a bit too hard for the ice sheet to grind down. This mini peninsula creates what seems like a long pier dividing Grand Transverse Bay into almost identical eastern and western halves from about mid "pinkie" on down.

Island View Orchards, a fourth-generation cherry farm on this peninsula, has been in continuous operation since 1876. A Swedish immigrant, the founder of Island View Orchards, purchased 40 acres on what was a wooded peninsula that today is surrounded by nature preserves, lighthouses, breweries, vacations spots, and wineries. It's a perfect haven for cherry production. But he didn't know it at the time. He didn't understand the advantageous ecological conditions of having land on a peninsula whose climate was perfectly moderated by one of the largest fresh bodies of water on the planet. He planted apples and potatoes. Not that apples didn't fare well and still do today. But cherries … oh, boy! They would love the cool summers that Grand Traverse Bay was able to provide while also preventing crippling harsh winters experienced by the deep interior of mainland Michigan.

Finally, in 1925, about the same year my house in College Park was built, they planted Montmorency cherries, the tart variety that was in high demand to make the perfect cherry pie. Today, the farm is owned by a third-generation grandson and his wife. His two brothers help with the daily operations; the next generation of sons, daughters, nieces, and nephews—along with five or six local kids—are temporarily employed during the July cherry harvest season. A migrant Mexican family group is shared with other farms to help with the fall apple harvest. The farm has grown to 180 acres, a third of which is leased land. Seventy-five percent of the farm operation is in the Montmorency cherry. Over the years, the farm has also added the Balaton cherry, another sour variety that is much darker and used in cough syrup.[31]

More recently, the area has discovered that grapes also do very well in the moderating climate around Lake Michigan. Island View Orchards has added four varieties of grapes that now make up 5 percent of the farm operations. In fact, the invasion of grapes into the Cherry Capital has sprung up a whole new commerce and tourism industry in Traverse City. Left Foot Charlie, an urban winery and cider maker in downtown, buys the grapes (and apples) from Island View Orchards. They stimulate local tourism with their tastings and tours, and then ship product to bars and restaurants as far away as Chicago and New York City.

Hanz made his living as a stone mason. But he was diagnosed with multiple sclerosis the previous year—about the same time that he lost his mother to a car accident, his father to sickness, and his dog to epilepsy. His attitude was unbelievably positive. He knew that if he exercised, he could prevent the progression of the disease. He was 45 years old and was as lean and fit as I was from having pedaled across seven states. His new dog, a chocolate lab named Duchess, ran in front of the truck as we drove through the orchard.

When we got back to Hanz's house, we had to take her down to the waterfront, so that she could fetch sticks out of Lake Michigan. Hanz's boat was anchored off in the distance about 200 feet from the shoreline. The sun's light filtered through the clouds, and a stiff summer breeze blew into our faces and pushed our hair back in wavy patterns as we stood on the clear, sandy bottom of Traverse Bay. We could have been in one of those shampoo commercials. The wind created the choppy waves that splashed across Duchess' chest while we waited for her to bring our stick back. Although Duchess was swimming, it seemed as though the bay was shallow enough that we would have been able to walk all the way to the boat.

It was a peaceful, meditative, almost prayerful scene. The sun was starting to set, so we went back to the house where Hanz's wife had supper ready. She had made pork chops, fresh sweet corn, potatoes cooked in onions, a cucumber salad, and, of course, a fresh cherry pie.

IF IT WEREN'T FOR the large bodies of water, such as the inland freshwater seas known as the Great Lakes and the oceans of Earth absorbing the heat, the effects of a warming climate would be much more dramatic in terms of what we "feel."[32] You might think of these enormous bodies of water as the central heating, ventilation, and air conditioning units of the planet.

When we get big snow falls, you might say, "What global warming?"

In 2017, the Rocky Mountains got more snow in January than they normally do all year.[33] The same huge snow event of 2017 in the Rockies, though, happened to be in the form of rain and flooding in the valleys of California—so much so that it brought California's Central Valley out of a drought, which had been so severe that no one in this generation had ever experienced anything like it; ecological and statistical evidence shows this magnitude of drought had not been experienced by the last two generations either.[34] The ocean temperatures do not fluctuate from season to season like the air temperature does; they may fall a few degrees, but the temperature baseline has been shifted upwards a bit. So, when the oceans release some of their surface moisture, and if it happens to be winter, it falls as snow in the northern latitudes and in the higher elevations.

The heavy rains that occurred in the spring of 2017 just west of the Mississippi River received almost no media attention, but they produced record-breaking flows at 14 USGS monitoring stations on several tributaries to the Mississippi River. At four of these tributaries, the previous record had been set only 16 months earlier. The monitoring stations that broke the all-time record during the first week of May 2017 *and* 16 months previously (in parenthesis) were:

Illinois River near Watts, OK (Dec. 28, 2015)
Gasconade River at Jerome, MO (Dec. 30, 2015)
Gasconade River near Rich Fountain, MO (Dec. 30, 2015)
Meramec River near Eureka, MO (Dec. 30, 2015)

This data would be the "flashing red light." Since these are all-time records, we can assume they are at least 100-year floods and quite possibly 500-year floods. Statistically speaking, the chance of a 500-year flood occurring in any one year is 0.2 percent (one-fifth of 1 percent or 2 out of 1,000). The chance of flipping a coin and getting heads twice in a row is 25 percent (0.5 x 0.5). The chance of having a 500-year flood two years in a row (technically, this is in three different calendar years but within 24 months of each other) might be 0.04 percent (4 out of 10,000).

These are back-of-the-envelope statistics because we haven't taken into consideration all the other events happening around the globe. But we can deduce that the chances of these recent events happening is probably somewhere between these two ranges: 0.2 and 0.04 percent. It's an unusual event. The oceans are storing most of the excess heat within the biosphere, and then releasing huge volumes of surface evaporation due to the lowered activation energy or "threshold" energy; the temperature distance between the surface of the ocean and the temperature at which water molecules will turn into vapor or gas has narrowed. It's as though the water molecules have gotten easier to be plucked away into the clouds. They are like a ripe cherry or apple that is about to fall from the tree compared to a sour one.

In North America, these clouds of moisture start off the coast, generally blow east, and then dump their moisture when they hit a cold front. Voilà ... record floods two years in a row. The rural community of Pocahontas, Arkansas, was probably hit the hardest with satellite data estimating the flooding to be 6 miles wide![35]

If you want a true barometer of climate conditions, take a look at the world's glaciers, which do not grow and recede on the whims of daily weather patterns. They are the slow and steady ice machines. If the global climate is warming, they slowly recede; if it is cooling, they slowly advance. Regardless of what individual drought or rain or snow events are happening locally, the pace of the glaciers remains consistent.

It is no different than how a stock portfolio works on Wall Street. If you are in a bull market and equities are rising at a clip of, say, 8 percent a year, at the end of the year your portfolio will be worth 8 percent more. However, there may be days when you lose $1,000 and other days where you gain more than 8 percent of the portfolio's worth. It's all what they call "noise" because, when it's all said and done at the end of the year, if you had a portfolio that was worth $100,000 at the beginning of the year, it's going to be worth $108,000 on December 31—no matter what the daily fluctuations were throughout the year. And vice versa: No matter what the ups and downs were through the year, in a bear market, your portfolio is going to be worth less.

Likewise, the glaciers tell no lies. They don't care what the daily fluctuations are in your neighborhood; they simply reflect the overall trend that is occurring. In Glacier National Park, the glaciers are mere remnants of what they were when my maternal grandfather immigrated to America from Slovenia during World War I. In fact, I saw not one glacier—not even a small glimpse—on the climb up to Logan Pass. They have receded too high into the upper elevations in one generation that none are visible without a hike into the alpine backcountry.

In the summer of 2004, I met up with my parents in Calgary, Alberta, and we drove to the Columbia Icefields in Banff and Jasper National Parks along the continental divide that extends northward above Glacier National Park. The Columbia Icefield can best be described as a flat, plateau-like area that holds year-round ice and feeds six major glaciers.

The ice has formed from centuries of accumulated snow compacting on itself. It is the freezer of the Rockies. The Icefield is basically a frozen lake that has outlets in several different directions atop a major watershed divide; however, the outlets are slow-flowing glaciers instead of fast-flowing streams. Whether melting or growing, glaciers move by gravity, but just much, *much* slower than liquid water.

The depth of the ice in many places would bury the Eiffel Tower, and it covers 125 square miles. But that number is changing as we speak. Within the icefield, mountain peaks poke up, like the way candles rise above the white icing on a birthday cake. When you are driving along the Icefields Parkway in the seam between mountain ridges, you cannot see the Icefield itself; you would need to be in a helicopter and look down on it. But you can catch glimpses of the toes of glaciers poking out.

One of these glaciers—the Athabasca Glacier—nearly reaches the Icefields Parkway, the toe of which is within walking distance from the Icefield Information Centre. You can park there and actually walk on top of it. In 1870, the Athabasca Glacier's toe would have poked into the Icefield Information Centre's lobby and parking lot. Today it is a 1-mile walk to it, and the walk gets farther every year.

Another interesting feature in Banff National Park is the Crowfoot Glacier. It was named by early French fur trade explorers because, from the seam above where the Icefields Parkway now runs, there were three toes that were visible where the glacier spilled down between small ridges and resembled the spread-out claws of a bird. At one time, it was fed from above by a smaller plateau-like area, separated from the Columbia Icefield but nearby, called the Wapta Icefield. Today, the glacier is isolated and completely cut off from the shrinking Wapta Icefield. One of the "claws" has receded away from view, and it no longer resembles the image for which it was named.

WHEN I LEFT TRAVERSE CITY, Michigan, the character of the trip had changed. I felt lonely and isolated. I couldn't figure out a way to get anywhere without zig-zagging on county roads that didn't head directly south. It was a complete maze of roads, and my head was spinning—Silver Lake Road, then the M-22, then right, left, right, left around pocket lakes, and diversions on cut-throughs around the mirage of nooks and crannies that is the shoreline of Lake Michigan. This route not only created navigation frustration, but it meant that I had almost no hope of coming across another riding companion as it would be impossible for these small, rural roads to funnel and concentrate the few cross-country cyclists out there as a major corridor would.

I would think, "Who else is crisscrossing northwest mainland Michigan by bicycle? Would they even be on the same roads that I am choosing?"

Finally, I found a freeway at Ludington, Michigan, and a policeman soon pulled me over. After looking over my driver's license, he said, "Son, I don't know what you people do in Alabama, but here in Michigan, a bicycler … cannot … be … on the freeway!"

"Can I go down to the next exit and get off?"

"No, sir. There is an overpass right there," he said, pointing at a hill leading up to a bridge over the freeway. "You need to leave now."

"Yes, sir."

I removed the panniers from the bicycle and carried them up the hill to the bridge overpass above the freeway. Then I walked back down and strained to push my bicycle through tall grass and around scraggly, unmaintained shrubs and saplings and up to the road. The cop watched me, as though I might disobey his orders. From up above and looking down at him, I watched him nod approvingly. He got in his car and drove off.

The cop had every right to kick me off the freeway because it is what Michigan law says. But it was like I didn't belong there, and it gave me pause for a second. Now my head was swirling with emotions: I was angry and sad at the same time. I wanted to be home. I was sorely disappointed and shocked that I couldn't ride on a nice, wide shoulder again. The irony was the fact that I was actually safer on the freeway with the 5-foot-wide shoulder than I was with cars whizzing past my outside pedal and left elbow on a two-lane county road with no shoulder. On my very first day way back in Washington, in the urbanized area around Everett, US Highway 2 had been a limited-access divided freeway, much like the one I had just been kicked off, but Washington State law did not prohibit me from being there.

Two days later, I crossed into my eighth state—Indiana—where the roads ran in a more familiar, more direct, grid-like pattern through farmland. I was determined to cross three more state lines.

Hanz and Duchess looking out over Grand Traverse Bay

8 Below the Divide

"Remember, Red. Hope is a good thing, maybe the best of things, and no good thing ever dies."—Andy Dufresne, *The Shawshank Redemption*

I ARRIVED IN COLUMBUS, Indiana, on August 17. But, before I reached Columbus, several people had offered me room and board; I took a few of them up on the offer. I was running low on funds, and the companionship was refreshing. The ride became lonelier after I had turned south. When I was paralleling the Canadian border, I saw riders in both directions every day. But now, other riders were far and few between. Along the Hi-Line, people riding bicycles were on a mission, going long distances to get somewhere. And the road was straight and wide with a clean shoulder. After I had turned south at the Straits of Mackinac, they were as likely to be homeless and wandering without purpose. I was as likely to get a "rebel" yell and a beer can thrown at me as I was to get a thumbs-up. It was a mix of despondency on my part, contempt on the part of drivers, and good-hearted hospitality from the local communities along the way.

One night in a public campground near Albion, Indiana, I was camping near a group of Mormons. They came in while I was cooking supper. I had just set up my tent and had not yet paid for my spot when the ranger came by and asked if I was with their group.

The lady in charge jumped in and emphatically said, "Yes! He is with us."

I became sick at one point and had no energy to pedal for three days. A man named Joginder—who came to the United States from Uganda, Africa, when the narcissistic Idi Amin took over his country—took me in and fed me three meals a day until I could continue my trip. His compassion almost makes me cry as I write this. I was ready to give up, and this total stranger from a continent thousands of miles away and a place I have only read about in news stories gave me what I really needed most: hope.

Joginder nursed me back to health (standing with one of his boxing students).

I had pulled into Marquette, Michigan, looking for a place to rest for a few days and maybe go see a doctor. I asked someone if there was a hostel in town.

"No," they replied, "but North Michigan University is out of session, and they rent dorm rooms for $17 a night during the summer."

I don't remember now how I met Joginder, but he was there, coaching at a boxing camp for Olympic hopefuls. He came to my door and knocked on it three times a day to make sure I was eating. He took me to the cafeteria and absolutely refused to let me pay for my own meals.

Back in Montana, people in the isolated prairie towns had been happy to see us and had pointed us to the city park to set up our tents and get hot water from the bathroom; many even had showers installed in the bathrooms. One day in North Dakota, bikers heading west asked me to deliver a postcard to a friend whom they knew in Ely. At Bay Furnace Campground in the UP, a couple from Stockton, California, had made me breakfast before I left. At Hog Island Campground, another couple had offered me supper while I was fixing the spoke in the rear wheel. South of the Mackinac Bridge, it seemed as though people were a tad leerier of strangers. One night in a rural area of Indiana, a state park ranger on midnight patrol shone a flashlight through the tent mesh and told me that I was camping in an illegal spot. Even though he was kind enough to let me go freely, without recourse, and even helped move me to a group camping spot, it was disconcerting to be awoken and put in a groggy confusion.

On the way to the proper camping area, we met a man named Berman, whom I will always remember by his firm handshake. He was from the local area but was camping at the park with a church retreat. He knew the park ranger and struck up a conversation with him while I stood waiting.

Finally, Berman looked at me and said, "Why don't you just stay under the awning of my camper instead of trying to reset your tent?"

The ranger looked at me and nodded his head in agreement. The next morning, Berman gave me coffee and breakfast and asked me if I needed money before I left.

In Columbus, Indiana, I met a kind, gracious, family man named Emanuele at the local Catholic Church. He took me to his house, so that I could get a shower, and then out to dinner. Emanuele was an Italian born in Holland and raised both in the Netherlands and in Indianapolis. We talked about all sorts of topics including the meaning of success, how to raise children, and the ideals of American society. We met some of his friends at the café, and he would often use the word "amigo" in reference to the person to whom he was speaking. For him, though, our relationship would be transient, and there would be no sentimental goodbyes. It was just two people crossing paths serendipitously, nothing more. No thought of sending a Christmas card or anything too committal like that. He was simply doing his duty to help a visitor in his own hometown. Although we could go into deep discussions about our personal lives, the relationship would end forever the moment I pedaled away.

I had left the Great Lakes watershed by crossing over the St. Lawrence River Divide somewhere along State Road 9 in northern Indiana a few days before; I had left the shore of Lake Michigan a few days before that. I had crossed the divide at 955 feet elevation. From the shore of Lake Michigan to where the St. Lawrence Divide crosses Indiana State Road 9, it is 125 miles as the crow flies. I had been gaining about 3 feet in elevation for every mile I pedaled, but I never noticed. From the time I crossed the divide in northern Indiana to Columbus, I had thus been on a downhill ride at about the same pace in reverse, but I never noticed it either; neither the climb nor the descent appreciably affected how fast I rode, how many miles I traveled in a day, or how hungry I was.

THE ICE SHEET THAT sat on top of North America during the height of the last ice age probably extended all the way down to about where Columbus, Indiana, sits today. It definitely would have covered all of the present-day Great Lakes and most certainly would have extended down to at least the St. Lawrence Divide. It most likely poked its nose—the very terminal part that poured out melt water—somewhere between the divide and where the Ohio River Valley sits today. The sheet of ice would have been hundreds or even thousands of feet thick, plowed down everything in its path, and gouged the earth as it advanced, the weight of it being unimaginable. The ice, being in a continuous, slow movement even once it reached its furthest extent, would have stopped growing because of warmer temperatures in the southern latitudes melting the nose away. Imagine feeding an endless Popsicle to a child who keeps licking the end, so that the length of the stick of ice never changes.

When the climate warmed and the ice sheet retreated, rudimentary rivers were formed on the top of the granite bedrock crust of Earth to carry away the meltwater that came in increasing amounts. This crust is many miles thick and, where the Great Lakes sit today, the crust was made of a much softer sedimentary rock. So, when the ice sheet retreated to where it now sits at the North Pole, deep and massive pockets had been gouged out that immediately filled with the fresh meltwater from the Wisconsinan Glacier—the part of the ice sheet that existed in what we know today as the Great Lakes region.

It took 100,000 years for the ice sheet to reach its maximum extent. The time chart that scientists have created, depicting the growth of this ice sheet, looks similar to a trending stock on the Dow Jones Industrial Average. Over time, it goes up a little and down a little, but the upward gains outpace the downward losses. Looking from a distance, over a long period, the stock has a definite overall trend; eventually it reaches a plateau and no longer grows in value, and then begins to decline. Although the ice sheet similarly slowly grew—two steps forward, one step back, and so forth—for tens of thousands of years, it only took about 10 centuries for the ice to retreat far enough back to reveal Lake Michigan.[36]

IT WOULD BE AMISS to call Columbus, Indiana, the "architectural capital of the world." It would be too pretentious for such a "young" town to try to compete with the iconic European cities or even other ancient cities on other continents. But it might be called the "architectural inspiration of the world."

As one visitor put it, "Even the bus stops are cool."

How this nondescript prairie town, squeezed between the White River and its tiny tributary called Haw Creek, reached worldwide notoriety can be attributed to the vision of Joseph Irwin Miller, the man who had to figure out a way to attract highly skilled employees for his new company, the Cummins Engine Company, during the post-World War II boom. Miller established the Cummins Foundation to subsidize architectural design a little snazzier than the run-of-the-mill school buildings that stoic Midwesterners might try to build. Before long, other types of projects received special funding too, and today Columbus is known as the "Athens of the Prairie."[37]

The next day, I crossed the Ohio River and into Louisville, Kentucky. To be honest with myself, I was tired of riding. Even the rural state routes through the UP had adequate shoulders, but the roads here had almost none, and it was mentally exhausting to keep up with the traffic behind me while constantly glancing over my shoulder. It was more like a sideways glance with my eyes while trying *not* to move my head too much, and getting as much information as possible from my peripheral vision. Your peripheral vision is always working, but most of the time you are not conscious of it. So, there was a continuous, conscious effort to keep my peripheral vision "turned on." It took a lot of effort, and I could almost feel the extra calories being burned by my brain.

I reached my home state on August 25. I had done 285 miles in the last three days. It was so hot and humid that I had salt residues piled on my skin after each day. Bent over with my head buried in my shirt and my arms flailing in front of me, I peeled off my shirt before going to bed. After leaving

Louisville, I spent one night at Beaver Creek Campground on Barren River Lake just outside Glasgow, Kentucky. I spent one night in Tennessee at Cedars of Lebanon State Park, about halfway across the state, and then crossed the Alabama state line the next day.

Three gentlemen (all brothers) in Kentucky who helped me fix a flat tire.

Physically, I could have gone on forever, but I was mentally drained. The narrow roads had pushed me to my limits, especially trying to navigate them on a loaded-down bicycle that was slow to turn and slow to stop, as though I were steering a battleship. I had been breaking a spoke in the rear wheel once a day. I had extras taped to the frame and had been replacing them every evening at camp. My dad drove up from Birmingham and picked me up. I was out of gas, and my trip was over.

I had been over two mountain ranges, through three time zones, and through four continental watersheds, and had ridden 3,221 miles. In an automobile, I could have easily completed this distance in less than a week; in a plane, it would have taken a few hours. But I had used only 6 percent of the energy requirements of an efficient passenger car with a combustible engine that burns fossil and probably about 3 percent of the energy requirements of an EV.[38] On an energy scale, the bicycle is an incredible bargain. It is the perfect balance between time and energy. If I had walked the distance I had just traveled, I would have nearly doubled my calorie burn, and it would have taken over half a year to do it. I had done it in 60 days.

MY MOTHER CALLED to say that my dad wasn't doing well and that I really needed to get over to the hospital. I left work early and walked to the bus stop to catch the local 10 to the train station. I got off the train at Lindberg and hopped the number 6 to Emory Hospital. On the way, I got a text from my aunt that dad had been moved down to the fifth floor to hospice care.

I was feeling a lot of anxiety. It didn't help that the bus driver was new to the route and was reading directions from a piece of paper while driving. I had been on the route once or twice before but couldn't remember if we were on the right road.

I would catch a familiar site out of the window and say to myself, "We're good."

As we were approaching the left-hand turn lane onto Clifton Road, the bus driver was staying in the right lane.

My anxiety grew and grew until, at the last possible moment, someone on the bus yelled, "Left!"

The driver looked in his mirror, swerved left, and asked, "Is everybody all right?"

Dad did make it through that day, and every day after work I made the same journey to the hospital. The next day, I was again seated on the number 6 behind the bus driver, and I picked up a small bit of conversation she had with a patron who was getting on at a bus stop in mid route. From her answers, I sensed that the bus driver was new to the route. I moved over to the right side of the bus, so that I could look straight out the windshield. As she approached the left-turn lane onto Clifton Road, she was staying in the right lane a bit too long.

Finally, I said, "You need to take a left."

"What?"

"You need to take a left."

"Here?"

"Yes!" I shouted.

She looked in her mirror, swerved left, cut off a car entering the turn lane, and stopped in the middle of the intersection. When it was clear that no one would be hit, she continued.

Vehicular traffic in the United States creates 34 times more emissions than every single bus running across the entire country. Vehicular traffic in Georgia creates *11* times the amount of emissions than all the jet airplanes taking off from the airport 1 mile south of my backyard, the busiest in the world—1,000 daily take-offs and two planes landing at the same time on parallel runways every 90 seconds for 12 hours straight.[39]

Let me say that another way: The sheer volume of us riding on top of our personal tailpipes overwhelms jet airplanes and big buses combined. Even airplanes are a form of public transportation when you compare them to the inefficiency of moving a person around in a box on four wheels. The reason the emissions difference between Georgia drivers and HJAIA flights are greater than between vehicles and airplanes for the whole United States (six times more) is because commercial airplanes, though they no doubt burn a lot of jet fuel, look efficient when you compare them to everyone in their car. So, if I don't get out and walk, ride, or use a bus, who is?

IT MIGHT HAVE BEEN surprising enough that the Clayton County Commissioner Chambers was standing room only. But what was really surprising was the fact that it was standing room only on a Saturday morning ... on the day after the 4th of July holiday. The topic of discussion was whether or not to allow the citizens of Clayton County to vote on whether or not the county would join MARTA as a full-fledged member.

Let me repeat the topic of discussion on that July 5th holiday weekend that needed to be decided in an emergency meeting: It was not to *join* MARTA; it was to simply *allow* a referendum on the issue in the upcoming November general elections.

There were four commissioners at the bench that day, and Jay Tifton, the chair, had been leading the charge ever since the state legislature passed the bill to allow local municipalities to raise their sales tax one penny if they used it to fund public transportation. The glitch was that the referendum had to take place during a general gubernatorial election season, and this was a general election year. And the other glitch was that the ability to raise the sales tax expired after this year. So it was now or never. One commissioner was absent, but her constituents were there in large numbers and must have taken notice.

There had been three public hearings on the issue since March. And those discussions seemed to have been dominated by one commissioner, Mitch Eagleton, who was there that day. He wanted, at most, one-half penny sales tax to be used for MARTA. In exchange, there would be no hope for any fixed transit such as rail.

He even pulled me to the side in the parking lot one day after a public hearing and whispered, "There will be a snowball's chance in hell the day Clayton County joins MARTA for a whole penny."

In the beginning, he got his way: The commissioners had approved the referendum for a one-half penny sales tax at the June regular meeting.

The MARTA board had immediately convened an emergency meeting the day after and said, "No. Even if the referendum passes, we won't accept Clayton County as a member for one-half penny. You would be tying into a system that the other members, Fulton and DeKalb counties, have been contributing with one-penny sales tax revenue for the past 40 years to build and operate a robust system already in place with 110 bus routes, 48 miles of rail, and 38 stations."

The haggling continued that Saturday with everything in the balance. Whether or not to place the issue on the November referendum had to be done within 120 days of the general election, so it was that day or never. Eagleton would not budge, and one other commissioner was firmly with Chairman Tifton; it was essentially deadlocked 2-2.

At the "11th hour," just before the meeting was to adjourn and after a lot of haggling, one commissioner changed her mind and declared, "I'll vote yes," and swung the vote 3-1.

The room burst into a roar and chaos; grown men and women were crying. The referendum passed that November by a crushing 78-22 margin. Needless to say, the one commissioner who was not there lost her seat two years later, and Mitch Eagleton lost his two years after that. And I could now catch the bus to Forest Park for the first time without worrying whether or not I had some extra change in my pocket.

TWENTY-SIX YEARS AFTER crossing the Alabama line on my bicycle, I was firmly rooted in my suburban community of College Park, Georgia. But are we really suburban? By the standards of population density, we certainly are. However, being next to the busiest airport in the world, there is a lot of land below the flight paths that has no residents, and that fact skews the true density.

By the standard of distance from the center of the urban core (7 miles) of the ninth largest metro area in the United States, we *could* be suburban; when the town was incorporated in 1891, it certainly was. On the other hand, the urban core is only five metro stops away, and the community has no less than six metropolitan transit buses zigzagging through the community. If you were to sit through a funeral service at the Episcopal church on East Main Street, you might notice a slight vibration in your seat every four and a half minutes as the metro train rumbled by at 60 miles an hour.

You'd think, "I must be in the city."

But, if I were to drop you inside Ward 4 where there are brick ranches on half-acre lots, you would say, "No, this is the suburbs."

If I dropped you off in the residential areas around the old depot where early 20th century bungalows sit on 50-foot-wide lots near four-story upscale apartments, you would say, "No, this is urbane."

If you saw the Family Dollar store sitting on top of an expansive parking lot along Virginia Avenue, you might be inclined to lean towards suburbia.

After you walked from that parking lot and then down the sidewalk for six blocks to a café on Main Street, you'd say, "Nope, it's the city."

Regardless, I had lived here long enough to gain enough institutional knowledge of the place that I could quote figures out of the $112 million annual city budget. I knew the balance on every one of the 13 revenue bonds our city owed. I knew the city manager on a first-name basis and most of the city's department heads. I also knew the homeless people by name who lived along the railroad line.

About this time, I got caught up in a battle with the governing body of the city over an issue whereby they wanted to sell another $36 million bond—the 14th one—and pay $9 million out of reserves as a down payment to build a professional sports arena for the minor league team affiliated with the Atlanta Hawks. I hired an attorney because the notice for the court proceedings to validate the bond sale said, in part, "Any citizen … in the City of College Park … who has a right to object may intervene and become a party to these proceedings."

And so we did. On the appointed day of the proceedings in Fulton County Superior Court, with the honorable Deputy Chief John J. Groger presiding, our plan was to object on the grounds that a municipality can't go into debt for recreational facilities that the community doesn't use. Seven citizens and one councilman were present.

When I sat down in the courtroom, Councilman Carver said, "John, there is $150,000 worth of attorneys in here representing the city!"

It was hard to tell who was whom. But, as I looked around the courtroom, there seemed to be about six attorneys to help represent the city. The city must have been a bit worried. Then, about 5 minutes before the start of the hearing, one of them walked over to my attorney and handed him a surety bond for $5 million. A surety bond was designed under Georgia statute to prevent a person from filing a frivolous lawsuit. Georgia law says that, if a person files a suit against someone, the defendant has a right to claim attorney fees and other damages, such as lost time and wages, from having to defend themselves if they win. The surety bond was a way to hold the plaintiff accountable and let him know that he would be liable if he lost the case—and let him know that fact *before* the proceedings began.

The problem is that I wasn't filing a lawsuit. I was simply becoming "a party to the proceedings." They were the ones who had filed for the validation. They were so scared of me and my one attorney that they had come up with a plan to circumvent basic democratic principles at the local level. It would have been akin to suppressing someone's right to free speech. To me, it felt akin to a book burning. Even more than that, I knew these people—the mayor, all the councilmen, and all the department heads—and had known them for years. You can't hold a public hearing? If you have the city's best interest at heart, shouldn't you be able to defend a $36 million publicly held bond? The judge let the proceedings go forth, and I was called to the witness stand.

"Mr. Duke, here are your Facebook posts. Aren't you really on a political campaign to smear the mayor?" asked one of the attorneys representing the city.

"Madam, I am fighting this bad deal on two fronts: It is both a financial and political battle. They cannot be separated."

She then asked me, "Mr. Duke, how much did you pay your attorney?"

I looked past the city attorney who was standing in front of me, then directly at my attorney who was sitting at the table, then at the judge, and then back at my attorney, looking for guidance. My attorney nodded in affirmation, but it still seemed to be an irrelevant question.

I turned back to the judge and asked, "Do I have to answer that question?"

The judge reprimanded me by saying, "You are to answer questions, not ask them."

"Fifteen hundred dollars," I replied.

In the end, the surety bond became a moot point.

Even though we didn't stop our city from going over the $200 million mark in total indebtedness over a development that had been tailored for a bunch of guys in Ralph Lauren three-piece suits, who lived God-knows-where and who looked at our community as a cash cow, we built quite a bit of political capital of our own. One incumbent councilman was ousted the year it was approved, and the mayor lost his reelection campaign two years later (four-year staggered terms).

Here is how I worded my thoughts at a council meeting 14 months before filing the objection:

"I am sure you have heard of the Paris Climate Accords. Mayors from many cities from around the world were there. While they were discussing solutions to this enormous challenge, there was another meeting next door called the Paris Accords to Build More Professional Sports Venues with Other People's Money. You might have been there. You don't even know what it is going to cost to build a venue for a wealthy basketball organization known as the Atlanta Hawks. Yet, you have made a blind commitment. According to the city manager, it would cost between $10 million and $40 million. That is like saying Wayfield has milk for sale for somewhere between $1 and $4 a gallon. I have seen this council agonize over whether or not to fund a part-time, $25,000 city staff position. But now you throw around these million-dollar figures as though you might be Warren Buffett or Bill Gates. But you are not. It is our money. To build this arena, you will have to: 1) deplete city savings; 2) divert tax revenues that were being used somewhere else; or 3) sell bonds and be in debt."

All three happened in the end. The final tab on the project came to $50 million. And, although city property tax revenues sky rocketed, three years later the city had to refinance all their debt for principal payment only, delay accumulating interest for three years, and push mortgage payments out nine years further because they couldn't balance the budget otherwise. I then went on to summarize:

"But the arena project is about status and another way of 'keeping up with the Joneses.' It is a project that creates a boondoggle that we have to maintain forever. The Atlanta Hawks will contemptuously call us 'sucker' underneath their breaths for funding their project. No other business entity gets this type of corporate welfare. If they want to have an arena in College Park, then let them buy the land and build it themselves as a private endeavor just like everybody else has to do."

I DON'T KNOW IF one can fully understand the importance of protecting the public domain for the general welfare of all, know the lives that other people live, and see the daunting environmental challenges that we impact in our everyday choices without getting out of your car; without riding a bus once in a while; without seeing how other communities do things, handle problems, and manage resources; without feeling the wind blow against your face; without following through on your convictions; and, quite frankly, without leaving the suburbs.

You can't have everything. You can't build a professional sports arena without something else being sacrificed. In my own community, we spent millions to build the arena, and there are cracked-up sidewalks, prominent bus stops with no shelters, and debris on the streets that seems to sit too long. If I don't object to bond debt that doesn't fix these problems, who will?

As I write this, the amount of CO_2 in the atmosphere went over a new threshold. On top of the Mauna Loa volcano in Hawaii sits an observatory run by the USGS. They have been measuring atmospheric CO_2 levels there every month since

March 1958. On that very first reading, atmospheric CO_2 was 315 parts per million (ppm); in March 2015, it crossed the 400 mark[40]; and, in 2021, it passed by 415. When the Wisconsinan Glacier sat on top of the Great Lakes, atmospheric CO_2 hovered around 180 ppm and it took about *7,000* years for CO_2 to rise to Earth's most upper "baseline" level of about 290 ppm while the ice sheet was pulling back to the North Pole.[41] We have raised it just as much in *100* years.

Earth breathes in and out, so to speak, on two scales: one yearly and one centamillennially (every 100,000 years). On a yearly basis, Earth's atmospheric CO_2 reaches a peak in May, and then begins to decline as spring foliage in the northern hemisphere reaches full bloom and plants are photosynthesizing at maximum efficiency.

It seems that, since the southern atmosphere—whose seasons are opposite from those in the northern atmosphere—has lost its foliage by May, there would be no difference whenever you measured atmospheric CO_2. However, the southern hemisphere is dominated by oceans, and the sheer volume of plant biomass in the northern hemisphere compared to the southern hemisphere drives the annual cycle of fluctuating CO_2.[42] Earth is "inhaling" from May to November and "exhaling" from November to May. But the total accumulation of atmospheric CO_2 has trended upward, like a credit card that never quite gets paid off and continues to accumulate charges.

Earth also breathes on a grand scale in which the inhale and exhale occurs over a period of about 1000 centuries. It's as though the yearly fluctuations are shallow breaths, and then there is a big sigh every once in a while. The direct driver of this long cycle has to do with a slight wobble—the Milankovitch wobble—in Earth's rotation around the sun. The Milankovitch wobble is like a bicycle wheel that wavers from side to side because it is missing a spoke. This wob-

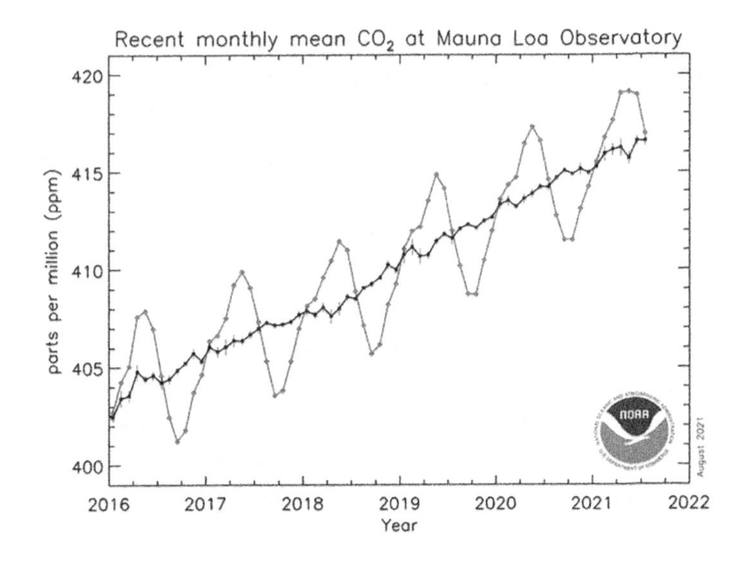

Image provided by NOAA ESRL Global Monitoring Division, Boulder, CO

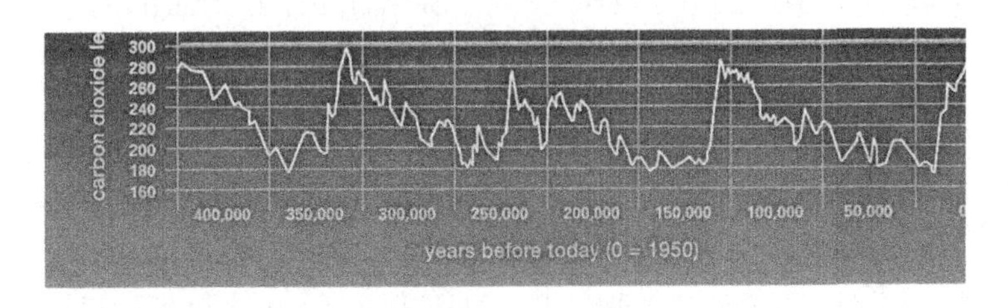

Courtesy NASA/JPL-Caltech

ble sets off a chain reaction of events that correlates with the slow fall in atmospheric CO_2 over tens of millennia, down to about 180 ppm as Earth cools; then, as Earth wobbles back, seems to cause a relatively sudden rise (7,000 years) back to what has consistently been a peak of about 300 ppm. This pattern is repeated over and over. The rising CO_2 may not be what *directly* caused the Wisconsinan Glacier to *begin* to back away from Columbus, Indiana, but it most likely amplified and accelerated its retreat back to the North Pole in the end.[43]

The question we should be asking ourselves is, "Where is the current upward trend in atmospheric CO_2 going to end?"

We are adding a tiny fraction every day. Take a look at one of those debt clocks and see the numbers twirling. With CO_2 we added 2.4 ppm in 2017 and about 23 ppm over the previous decade since 2017. During the decade of the 1960s, we added about 8 ppm; in the 1970s, we added about 13 ppm; the 1980s, it was about 16; the 1990s, it was only about 15; and the 2000s, it was about 19. The overall trend seems to be generally *accelerating* at about 3 ppm every decade, and we will hit the 500 mark in the year 2058 if not sooner.

What is that going to be like? If ppm's were currency, and 1 ppm were a $100 bill, the debt clock would be "spinning" at 65 cents a day. That may not sound like a lot, but if Earth itself were measured in ppm relative to our solar system, its mass would only be 3 ppm. To put it another way, you could be charged with driving under the influence if the amount of alcohol in your bloodstream were 800 ppm.[44]

It is incredible to think that we do have the power to tweak the climate. But what is also intriguing to me is the fact that, if indeed birth is a random event, the odds of me being born in *this* wealthy continent is about 1 in 20.[45] With that in mind, how could I ever have complained about any of the hardships I experienced on this long journey? The heat and cold, the wind and rain—I could not have fathomed fretting about such things. To do so would have only aggravated any troubles that I might have encountered.[46] It also would make me seem insensitive, unaware, and conceited. The fact of the matter is that the vast majority of the world lives on less than the equivalent of $10 (US) a day.[47] They spend every waking moment just fulfilling the basic needs of life. Poverty in the United States runs at about 13 percent, and the poorest adults in the United States—about 24 million—work and live on a minimum wage in which housing consumes most of their income.[48] If the other four out of five people in the world were to climb the social ladder to my middle-class status, including those in my own country *and* in the developing world, the environmental impact would be the equivalent of increasing the world's population 12 times over.[49]

They, the poor, are the ones who are riding the bus, walking, and riding a bicycle as fundamental forms of transportation to really get somewhere and go about their day. They have the smallest environmental footprint of us all. If not for the poor among us, there would be even less of a public transportation network; asthma rates from ground-level ozone would be even higher and the roads even more congested. And a rapidly warming climate would be more daunting than it already is. So, why do we subconsciously and culturally view those standing at a bus stop in the rain or walking on a worn, dirt path along the side of the road as second-class citizens just because we haven't yet finished investing in basic infrastructure?

SIX MONTHS AFTER THE bond was approved to build the arena, I had rebuilt a decent enough relationship with the mayor, and he agreed to listen to some of my ideas. He would be up for reelection within the next 18 months when I sat down with him and asked for his support to finish the Brady Trail. The Brady Trail was a 10-foot-wide, multipurpose bicycle and pedestrian pathway that had never been completed and had sat at a dead end for over 10 years in the middle of a forested area next to the municipal golf course. The trail was to be finished along a tributary stream to the Chattahoochee River. The funds for the project would come from sales-tax revenues collected by the county and paid to the city solely for transportation projects. In late summer, the mayor sponsored the legislation, and the council approved it unanimously.

Less than a year earlier, I had stood in front of them and had defiantly pointed my finger at each individual on the governing body, saying, "By God, we will remind the citizens at the next election how you heavily subsidized a wealthy professional sports team."

Soon after, city budget meetings started, and I attended every one of them. At the very first one, I had sat in the small, cramped conference room no bigger than my kitchen for two hours and was ready to go home when the meeting was over.

Then, as everyone was collecting their things, the mayor said, "Before we go, I've got one other thing I want you to discuss."

Councilman Carver said, "You can talk about it all you want. I am a 'no' vote, and I am going home." And he got up and walked out the door.

I stayed seated, and the atmosphere got chaotic, to say the least. The mayor wanted to give a $250,000 tax break to an apartment building, which was charging $1,100+ a month in rent and had just been built in the downtown area near the MARTA station.

The mayor said matter-of-factly, "It's about $250 on each citizen of College Park."

The Ward 1 councilman was a meticulous engineer who read and analyzed every document word for word that ever came before the governing body. His ward included the most prominent homes on the premier residential street of Rugby Avenue where 200-year-old oaks line the street and included the historic building owned by the prestigious Woman's Club at the corner of Main Street.

The Ward 1 councilman had stayed seated and said, "You know my position on this. My answer is no." He argued that tax breaks are used as incentives to bring in development, but this project was already built.

The Ward 3 councilman was the same person who had opposed my support of the variance for the restaurant owner 12 years earlier. He tried to convince the Ward 1 councilman that the developer of the apartment building, who was asking for a handout, is promising more high-end development near the MARTA station in the near future.

The Ward 1 councilman seemed to feel as though he was being brow beaten and in exasperation declared, "Mayor, it just doesn't feel right."

I asked if I could speak, and the mayor immediately said, "No."

Meanwhile, the Ward 2 councilman, who had just ousted the incumbent over the arena deal and who had been a former city sanitation worker, was sitting quietly through his very first budget meeting. The mayor called for a break in the meeting and let the Ward 1 and Ward 3 councilmen go into the hallway and resolve their differences. While they were in the hall, I walked over to the stenographer and asked if the meeting was being recorded.

The mayor jumped in and said, "Yes, it's being recorded!"

I said, "Sir, I was speaking to the stenographer."

"It's not your place to run the meeting, but you are welcome to listen."

"There is no doubt about that."

"If you want to get crossways, I will have you removed!"

"For what?"

"For runnin' yore mouth."

I chuckled and went back to my seat. When the two councilmen came back into the room from their hallway sidebar, the mayor called for a vote. The Ward 1 councilman had suddenly and miraculously changed his mind, and the tax break was approved.

I did an open records request the very next day to obtain the two-hour audio, clipped out the 20-minute piece about the tax break, and was ready to post it on the neighborhood Facebook page. But, before I did that, I wanted the developer to give his side of the story. I really wanted the bike trail finished and had been at the meetings in the first place to see how I could get the funding. First, I went to see the mayor privately and without an appointment to let him know that there was nothing personal between us and that it was just business. He thanked me and shook my hand.

When the developer and I sat down at the local bar and grill, he knew something was up. We small-talked for a few minutes and then he asked, "What is it that you *really* want?"

"With all the things this city needs, and with a third of the population living below the poverty line, how do you justify receiving a $250,000 tax break?"

He seemed angry at my question and accused me of many things. When he saw that his accusations weren't going to stick, our conversation continued. About an hour into the discussion, I told him about my idea of finishing the Brady Trail. His whole demeanor turned 180 degrees. He was for that, offered help, and paid for my meal, but I never heard from him again.

In the end, I never posted the audio and, when it came time for the public hearing, I spoke about how during 12 hours of budget meetings I never got a sense of long-term goals. But I also praised the city for paying off one of the bonds on an enterprise in the city that now has positive cash flow. I never mentioned the fiasco at the first budget meeting.

When I sat down, the Ward 3 councilman said, "I have never seen a citizen sit through all the budget hearings. For someone to do that, I at least need to hear what their ideas are."

And the mayor seconded.

By Halloween of that year, the Brady Trail was finished and connected all the way to a sidewalk on a major north-south corridor running on the back side of the golf course. Studies done by Aerotropolis Atlanta—a self-taxing community improvement district around the airport—estimated that it would have cost $159,000 to finish the 700 feet of trail that I had been commissioned to do. I built it on a budget of $30,565—a mere 1 percent of the transportation sales tax revenue that the city receives in one year.[50]

Finally, those who lived on the west side of the city had a way to safely and efficiently get downtown and to City Hall, the library, the polling place, and the MARTA station in ways other than by driving a car.

9 The Way Forward

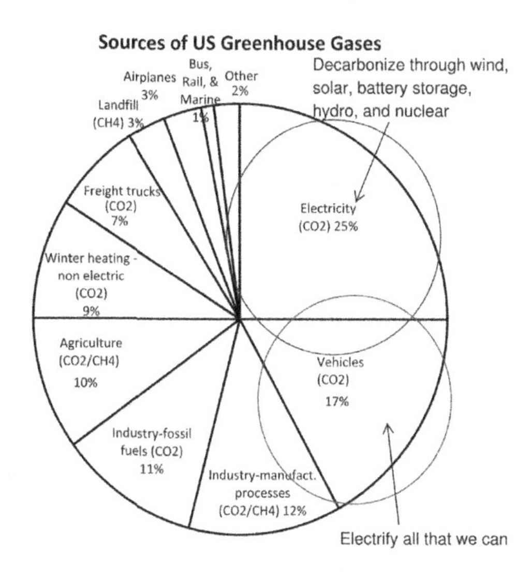

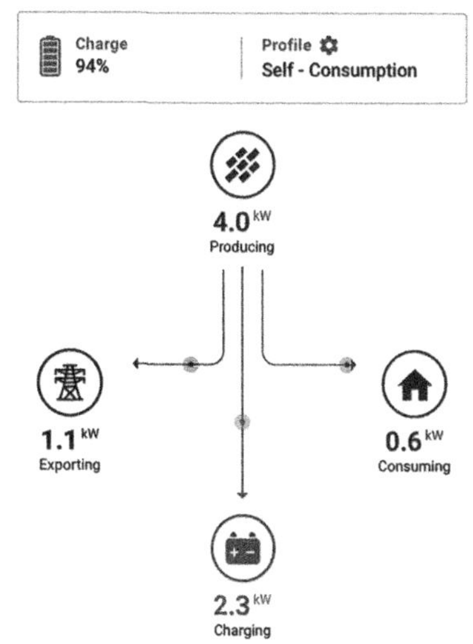

IN THE SUMMER OF 1996, emergency room visits in Atlanta due to childhood asthma suddenly fell by a third, while all other emergency room visits chugged along per usual.[51] Why the sudden drop in acute asthma cases? The Olympics were in town, and traffic controls had been put in place to accommodate the crush of visitors. Roads were not exactly empty, but many people used transit—some for the first time, if they went into work at all—as many residents simply rented their houses and left town for a while. This public health study showcases how clean the air could become, which is a different issue in itself (ground-level ozone) from the climate crisis, if we stopped burning fossil.

I don't think it is God ordained that we burn mined fossil for our energy needs. And this book isn't meant to change anyone's political view, especially anyone hell bent on denying the science. In fact, as mentioned in Chapter 1 about Martin Luther King's use of my childhood state to gain traction in the civil rights movement, it is quite the opposite.

This book is meant to offer a way forward for those who care and can grasp the enormity of this humanitarian challenge but who may have been duped by the fossil industry for so long that they quite possibly still (subconsciously) believe the decades-long propaganda campaign by the industry's public relations hired guns. Those of us on the front lines of the climate crisis are like the abolitionists in 1850. It looked daunting and no one believed it could happen—until it did.

The move to a carbon-free economy is already happening. Private corporations, entrepreneurs, utility companies, and everyday people like you and me are making the necessary changes and leading the way—not the politicians, as I have shown throughout this book. The children in my family decided to go vegetarian (not vegan) in order to reduce their carbon footprint. So, we learned to make one vegetarian meal for everyone to share at suppertime. We all share one vehicle, and I walk or ride the bus when the car isn't available. These are small, empowering, and meaningful steps.

But we can also absolutely leave fossil in the ground. As I pointed out in Chapter 6 and in the Introduction, renewable energy has now become a cheaper way to produce electricity than by continuing to run fossil fuel plants. Not cheaper than *building* fossil fuel plants but cheaper to build out renewable and actually *close down* fossil fuel plants. Xcel Energy announced in 2019 that it plans to shut down all its remaining coal power plants a decade ahead of schedule by increasing maintenance on existing nuclear facilities, by investing in 4,850 MWs of wind and solar farms, and, just as important, by promoting efficiency among their ratepayers.[52]

Xcel Energy is based in Minneapolis but serves 3.7 million customers in seven other states as well. Will this disrupt things for some stakeholders, like the local communities, who may depend on these plants for property tax revenue and jobs? Of course. But we have been through these transitions before and, with advanced notice, the communities are adapting through coal-power plant job attrition, job transfers, and retraining programs through community colleges, and economic redevelopment of the real estate.

These are important milestones. If we can decarbonize the grid and electrify everything else, we can cut US greenhouse-gas emissions by about 42 percent. If I had made this statement just five years ago, I would have said that it would cut US greenhouse-gas emissions by half.[53] But we have made progress in the electric grid, and the amount of greenhouse-gas inventories have shifted in the proportional pie chart, which does not show raw emission data but is used to exhibit the "low-hanging fruit" and the biggest emitters to cut, which will make the most impact.

Just five years ago, solar and wind power were still an expensive way to generate electricity. As solar and wind power have increased capacity, there has been a domino effect: With each doubling of capacity, cost has fallen 20 percent. In 10 years (between 2009 and 2019), the cost of wholesale electricity from a wind turbine fell off a cliff from 36 cents to 4 cents a kWh.[54]

Lithium ion battery storage prices have taken a similar course. As I pointed out in Chapter 6, batteries are so cheap now that we added one to our home solar system, and we now send less than 1 kWh to the grid per day. And that is mostly because the system is making (imperfect) decisions in real time and some "spills over." I now realize that having a solar system without a battery really does not make sense. It just takes a small bit of battery power to even out the solar supply and make it more efficient.

What all of this means is that I am now essentially a part of the grid, and the investment I made is no different than if I had invested it in a retirement portfolio. Either way, I am getting a return. In other words, it isn't just rich people who are doing these kinds of investments. It is us: the middle class. I am helping to "democratize" wholesale electricity, reducing peak power during hot days in the summer when the A/C is running so that a fossil plant can be shut down sooner, and forcing competition on wholesale electric rates, which benefits my neighbors as well.

Airplanes will probably be one of the last things to decarbonize in our economy, but as I pointed out in Chapter 2, Chapter 4, and Chapter 8, they are part of the public transportation sector that includes city buses and rail whose emissions are vastly overwhelmed by other sectors of the economy and our daily lives. What airports can do, meanwhile, is decarbonize their ground operations through electrification of baggage tugs and place solar panels on top of the terminals. Single-use consumption, however, is a massive cumulative impact that is being addressed right now.

Single-use consumption includes plastic eating utensils, disposable grocery bags, our morning coffee cup from Starbucks, or—in the case of the climate crisis—filling your gas tank. The cumulative impact of pumping gas, which I pointed out in Chapter 1, is what makes it such a huge chunk of the greenhouse-gas pie chart. Again, the private sector is taking the lead in electrification. General Motors, for example, has said that they will not be selling internal combustion engine (ICE) vehicles within 14 years.

It is incredibly liberating to drive an electric vehicle, and owning one feels like I am carrying a smart phone while everyone else is still walking around with a flip phone. There is almost no maintenance, it has fast acceleration, and it is much cheaper fueling it in the driveway. There is no "idle" and the battery recharges "on the fly" when you lift up on the accelerator. You don't "crank" the car; you push a button and go. There are no starter, fan belts to squeak, spark plugs, oil changes, emissions inspection, muffler, tailpipe, or transmission. There is maximum torque in the single-gear drive shaft from the start, which is why it can go from 0 to 40 in about 3 seconds, and why the car is driven with "one pedal" and, therefore, requires almost no braking maintenance either. You could take off four bolts and replace the entire electric motor and drive shaft in one fell swoop. With or without the climate crisis, electric vehicles are simpler to make and have superior technology that will drive down transportation costs for everyone.

General Motors offers an eight-year, 100,000-mile warranty for their all-electric vehicles, including guarantees on battery charge level. I rented one in Minneapolis that had

120,000 miles and drove it to Duluth on a single charge. I plug in our Chevy Bolt at night to make the grid more efficient. You can't just shut down a nuclear facility in the wee hours of the night when everyone is asleep, so I am charging the car with electricity that MEAG doesn't know what to do with. Therefore, I am helping to drive down electric rates for everyone. I put in my own car charger with $40 in parts and $180 for the charge cord, which draws less electricity (20-amp breaker) than the hot water heater (30-amp breaker).

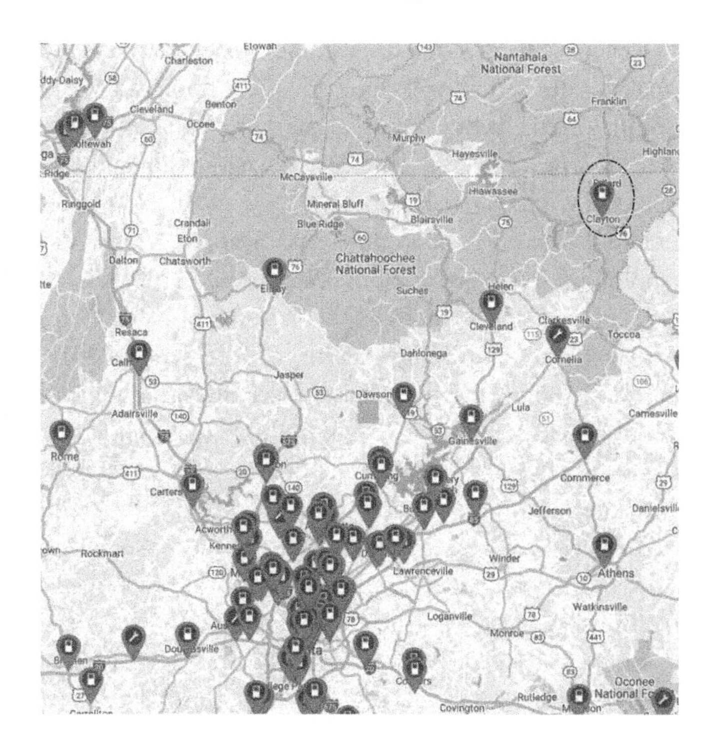

Pictured above the network of Level 3 fast chargers (50 to 150 kW) that already exists between my home and Great Smoky Mountains National Park. Although many chargers exist at Walmart Supercenters along the interstate freeways, I use the one circled in downtown rural Clayton, Georgia, where the charging is free because the community is promoting tourism as a gateway to the mountains. I get a meal there and watch it charge on my phone. I can get back home on a single charge because it is "downhill" from Gatlinburg, Tennessee, to Atlanta, Georgia. This map does not include the Level 2 chargers, similar to the one in my driveway (3.5 to 7 kW), which are located at hotels and other public areas like places of employment, shops, and restaurants. We don't use any of these Level 2 chargers in Atlanta because we never need them.

As electric vehicles build in volume, just like solar and wind capacity, their prices will come down. As I pointed out in the Introduction, the robust used-car market for the five-door, all-leather interior, 220-mile range Bolt has already pushed their price below $20,000. Have there been glitches? Indeed. Twelve battery fires in 141,000 Bolts sold over five years. As I tried to point out in Chapters 7 and 8, it is not natural for us to gauge risk and probability though. That number of battery fires equals 0.000085, or about 9 thousandths of 1 percent. In 2018, there were 144,500 vehicle fires due to mechanical failure (not crashes) out of 276.3 million US registered ICE (gas) vehicles.[55] That is about 0.00052, or about 5 hundredths of 1 percent. It is much riskier (18 times, in fact) to drive around with a tank of gas than it is to have a battery. (None of the battery fires occurred while driving; only sitting still and fully charged.) Those 12 Bolts would have driven about 4 billion miles if the average annual mileage on a car was 10,000. That is 0.3 battery fires per 100 million miles driven. To put that in perspective, there were 1.37 deaths per 100 million miles driven on U.S. roads last year.[56]

The electrification revolution through battery storage is creating an economic boom. A defunct Mitsubishi plant in Normal, Illinois, which was shuttered in 2015, has been revitalized by Rivian, maker of electric delivery vans, and now employees about 2,000 people.[57] The solar panels on my home were made by the Hanwha Q Cells factory in Dalton, Georgia. And, even though I have no problem with my Bolt, I might receive a new battery, paid for by General Motors, which was made by the SK Innovation lithium-ion battery factory in Jackson County, Georgia.

Do these manufacturing facilities carry a carbon impact? Yes, they do. If you look at the pie chart, probably about 23 percent of current emissions from *all* industries. But, if we can end the fossil infrastructure—mining, shipping, refining, and distribution—for fuel use, we can also cut industry emissions by about half (11 / 23) and reduce overall greenhouse-gas emissions significantly. Why have duplicate systems when we have to have an electric grid anyway? We can focus our investments to make it better. As everything else is electrified and the grid becomes carbon free, the other hard-to-get pieces of the proportional pie chart that will be the last to decarbonize will increase, but overall emissions will fall.

The car charger I installed at my mother's house for her Bolt.

Epilogue: The Drama of a Changing Landscape at Slow Speed

IN THE SUMMER of 2018, I went back to the area in northwestern Minnesota where the Great Plains ends and the forest around the headwaters of the Mississippi River begins. It had been such an emotional and profound transition for me on that summer bicycle ride that I had to clarify a few things from the journal I had kept.

I felt hypocritical to be driving a gas car along US Highway 2. I flew standby though, and waited for an open seat to get from Atlanta to Minneapolis and then a connection on to Grand Forks, North Dakota. I have sometimes been in airports all day waiting for an open seat. I essentially didn't affect the flight schedule, and my ticket—at least from the perspective of the airline—had no value. So, if my flying wasn't carbon neutral, it was at least carbon light.

Of course, it would be more efficient for me to have taken a passenger train or to have ridden my bike. But the longer the haul, the better a full commercial flight looks on efficiency.

It is 1,450 road miles from my house to Grand Forks. If I had driven there, I would have burned about 40 gallons of gasoline. Since I would have driven alone, that equates to 40 gallons of fuel per person. But let's say that my wife wanted to see where I had ridden too and jumped in the car with me. Now we are down to 20 gallons of fuel per person. The highest-capacity plane going from Atlanta to Minneapolis—a Boeing 757—holds 200 people, not counting the crew. Let us assume it lands in Minneapolis, lets people off, lets others board, and takes off again for Grand Forks.

It is 909 air miles to Minneapolis and 272 air miles on to Grand Forks for a total of 1,181 air miles to get to my destination. The Boeing 757 will burn 2,927 gallons of fuel to get there. That equates to about *15* gallons of jet fuel per passenger and a fuel efficiency of nearly 81 miles per gallon per seat. The Boeing 757 was designed in 1982. Fuel efficiency has improved since then, and the Boeing 737, designed in 2017, now gets *103* miles per gallon per seat.[58]

For *intra*-city travel, we can compare and contrast using a unit of measurement called "pounds of CO_2 per passenger mile," which also takes into account, like the Boeing 737, how many people are riding. In this case, however, you want the number to be as low as possible. If I ride the train in New York City, I am contributing 0.15 pound of CO_2 for every mile I ride. If I ride the train in Atlanta, I am contributing 0.25 pound of CO_2 per mile.[59] The Atlanta train is newer and more efficient, and therefore takes less electricity to run. But the power grid that supplies it (Georgia Power, not MEAG) gets about one-third of its power from carbon-free sources while New York City is getting about *60 percent* of its power carbon-free.[60] And the average train in New York is about 50 percent fuller than the average train in Atlanta. For comparison, the average US single-occupancy (passenger) vehicle produces 0.96 pound of CO_2 for every mile driven. The MARTA bus system produces 0.78 pound of CO_2 per passenger mile driven, and the average bus is only 21 percent full while the Metropolitan Transportation Authority New York City bus system is producing only 0.56 pound of CO_2 per passenger mile because the average bus is twice as full (41 percent).[61]

So, what bothered me most was the car rental. Although I had detoured to Ely on the bike at Grand Rapids, Minnesota, my plan was to drive the diagonal-cutting highway across the state on a continuous, straight shot to the Duluth, Minnesota airport, drop it off, and fly back home. Although we now own an affordable EV, which is charged about 95 percent of the time in our driveway, it is still very difficult to *rent* an EV. However, the charging network out there is already well established (check out the PlugShare app). Our Chevy Bolt gets 220 miles of range on a single charge and, although it is 263 miles to the Duluth airport, there is a Level 3 fast charger about halfway in Bemidji, Minnesota. No problem. It would have taken about 30 minutes to top off the car in Bemidji while I had gotten a meal, and I would have been able to drop off the car with over a quarter of a "tank" to spare.

I wound up renting a gas car at the Enterprise counter at the Grand Forks airport, drove 45 miles an hour, and turned on my flashers as cars zoomed past me in the passing lane. On the two-lane parts of US Highway 2 in Minnesota, I eased to the 5-foot shoulder, which I had ironically used on my bicycle, so that cars could easily get on by. This tactic seemed to work well, and it allowed me to take in the route about as slowly and safely as possible. I averaged 50 miles to the gallon in a Chevrolet Cruze and used less than 6 gallons of gas to do my research.

Here is what I found: Some places that I have written about in this book are no longer there. For example, I could not find the bar and grill where I had lunch and played the jukebox on the next day after crossing the Red River into Minnesota. Who knows about the rest of the route that I haven't revisited? I especially worry about those small places in North Dakota and Montana.

The United States has added 75 million people in the years since, and therefore the average density of the country has increased from 67 people per square mile to 86 per square mile. Lakota's population in North Dakota, on the other hand, has dropped from 892 to 694. Culbertson, Montana, has gained about one person per year since my ride through there. The café in Fisher, Minnesota, where I ate breakfast on the morning I left Grand Forks and just before I came to the first lake in Minnesota at the town of Erskine, was no longer there. A Cenex gas station on the corner of 4th Street and US Highway 2 was the only retail establishment left.

But I did find the café where I had eaten breakfast the morning I rode to Bemidji.

When I asked the waitress how long the place had been there, she said, "Since 1991."

I was a patron the year it first opened! Minnesota State Road 89 merges onto US Highway 2 at the exact location, and the lifeblood of a major rural intersection seems to keep the restaurant thriving. I also found the headwaters of the Mississippi River to be as clean and wild as I had found them before. I camped along the shore of Lake Lomond in the Bagley City Park among a grove of birch trees nearby, as I had done years before, hummed the melancholic lines to *The Bonnie Banks O'Loch Lomond*, and watched the sun set.

References and Notes

1 https://www.eia.gov/todayinenergy/detail.php?id=46416.

2 Those who live near facilities where recyclables are actually recycled and where the consumer buys products back make a significant contribution to the cause. For example, there is an aluminum recycler about 12 blocks from my home. There is a lot of discussion about this topic, but it is lacking in scientific research on the intricate math of recycling's effect on net greenhouse-gas emissions. The more specific truth is that aluminum is probably the only thing that has enough value to industry to truly spin through the cycle fast enough to make a significant impact on mitigating our consumptive behavior. Glass, for example, is so heavy to move around and so cheap to make from raw material that the economics of using recycled glass don't mesh with the public's overwhelming desire to reuse it; the carbon footprint from recycling glass is probably greater than it is from just making it from raw material. The largest receiver of our throw-away plastic used to be China. We would ship it to the coast, put it on a ship in Los Angeles, and sail it across the Pacific Ocean. China has stopped receiving plastic from the United States because the plastic had become so dirty—literally dirty, like with food scraps and such. That tells me that people are using recycling as a means to absolve their habit of overconsumption without thinking of ways to reduce it first. In the bigger picture, recycling is now a way for people to feel good about themselves while making daily choices that adversely impact the planet in a worse way. How big was that carbon footprint to recycle plastic in China and then have them ship cheap plastic goods back to us that we would toss aside again? Kimiko

de Freytas-Tamura, "Plastics pile up as China Refuses to take the West's Recycling," *New York Times,* January 2018, https://www.nytimes.com/2018/01/11/world/china-recyclables-ban.html; Aluminum Recycling Facts: http://www.recycling-revolution.com/recycling-facts.html.

3 US Department of the Interior, National Park Service, Glacier National Park: http://www.hikinginglacier.com/glacier-national-park-cycling.htm.

4 US Department of Energy, Energy Information Administration: http://www.eia.gov/petroleum/supply/weekly/.

5 There is an abundance of scientific work published on this topic (e.g., John M. Drake and Blaine D. Griffen, "Early Warning Signals of Extinction in Deteriorating Environments," *Nature* 467 (2010), 456-459).

6 John G. Mitchell, "Oil Field or Sanctuary?" *National Geographic,* August 2001, http://www.pollutionissues.com/A-Bo/Arctic-National-Wildlife-Refuge.html.

7 Joel K. Bourne, Jr., "This wilderness crown jewel is opening for oil drilling. Why is industry interest so weak?" *National Geographic,* January 2021, https://www.nationalgeographic.com/environment/article/wilderness-crown-jewel-opening-for-drilling-industry-interest-weak?loggedin=true.

8 There was a lot of reporting on this subject, including the fact that tankers were sitting off of the coast of California waiting to dock because there was nowhere to put the oil. Here is a no-spin reference from the Congressional Research Service website: https://www.everycrsreport.com/reports/IN11354.html.

9 Tax expenditures are those numerous items that we are allowed to deduct from our income. According to many economists and the Simpson-Bowles plan, they should be thrown out of the tax code, and Congress should be made to pass legislation making them a government expense that they have to figure out how to pay for. The "zero option" of the Simpson-Bowles plan strips down the tax code and simplifies it while reaching

a balanced budget within about a decade. The nonpartisan Committee for a Responsible Federal Budget has taken up the advocacy for getting Congress to at least pass a budget and to get deficit spending under control. http://www.crfb.org/blogs/zero-plan-re-emerges-tax-reform-debate.

10 The continental divide that runs through Logan's Pass is better described as the western continental divide because there is another continental divide in the Eastern United States that runs the ridge of the Appalachian Mountains. The Appalachians were formed at a much earlier time in Earth's history and are therefore eroded down to half the height of the Rockies. But height is not the only factor that determines watershed divides—it's also consistency. Nonetheless, the ridge of the Appalachians—known as the eastern continental divide—separates a different major ocean (the Atlantic) from the same river system (the Mississippi) as the western continental divide.

11 21,596 is half of 43,496. http://www.ncaa.org/championships/statistics/ncaa-football-attendance.

12 Ayn Rand, Leonard Peikoff, and Peter Schwartz, *The Voice of Reason: Essays in Objectivist Thought* (New York: New American Library, 1989).

13 GHG Inventory_BW_Grayscale_x.jpg, https://www.epa.gov/ghgemissions/inventory-us-greenhouse-gas-emissions-and-sinks-1990-2019.

14 See Endnote #25 for an explanation of a kWh. Here is the math for fuel comparison between electricity and gasoline: 34 kWh x .18 (18 cents) = $6. If the car gets 4 miles per kWh, then it can go 68 miles on the cost of a "gallon" equivalent if gas is $3 a gallon (4 miles x 34 kWh divided by 2). This distance is 2.7 times farther than the average combustible car engine on the road today (25 miles per gallon). So, $3 divided by 2.7 = $1.10.

15 Union of Concerned Scientists: https://blog.ucsusa.org/dave-reichmuth/plug-in-or-gas-up-while-driving-on-electricity-is-better-than-gasoline/; US EPA: https://www.epa.gov/egrid.

16 https://www.2degreesinstitute.org/reports/comparing_ghg_emissions_of_bevs_and_icevs.pdf.

17 Christophe McGlade and Paul Ekins, "The geographical distribution of fossil fuels unused when limiting global warming to 2 °C," *Nature* 517 (January 2015), 187-190.

18 Here are two sources of information on planetary atmospheres: University of Calgary (http://energyeducation.ca/encyclopedia/Greenhouse_effect_on_other_planets) and Space.com (online magazine) (https://www.space.com/18067-moon-atmosphere.html).

19 This link may change as it just came out at press time. "The Summary for Policymakers" is a fairly easy read with relatively simple graphics. https://www.ipcc.ch/report/ar6/wg1/.

20 "The 2001 Outbreak of Foot and Mouth Disease," National Audit Office (United Kingdom), HC 939 Session 2001-2002, 21 June 2002.

21 William Rannie, "The 1997 flood event in the Red River basin: Causes, assessment and damages," *Canadian Water Resources Journal* 41 (June 2016), 45-55; Dan Gunderson, "20 years after epic flood, Red River towns no longer dread the spring," Minnesota Public Radio, April 17, 2017.

22 Okpik (pronounced "ook-pick") is an Inuit word that means "Snowy Owl." There is actually a species of bird we call Snowy Owl (*Bubo scandiacus*). The Inuit are a Native American people who live a sustenance life in the Arctic. They have about 50 different words to describe what we call snow—whether it be a wet snow, a dry snow, a blowing snow, and so forth.

23 Great Lakes Information Network: https://www.glc.org/glin; US Environmental Protection Agency, Physical Features of the Great Lakes, https://www.epa.gov/greatlakes/physical-features-great-lakes.

24 Dan Gearino, "Plan to Save North Dakota Coal Plant Faces Intense Backlash from Minnesotans Who Would Pay for

It," *Inside Climate News* (July 23, 2021). You can check out the mapped coal plants still remaining in the world, how much electricity they produce, how old they are, and what their future status at https://www.carbonbrief.org/mapped-worlds-coal-power-plants.

25 A kW is the potential power production; kWh is the work actually done. It is like saying, "I can run 5 miles" versus "I can run 5 miles in 20 minutes," which takes more work than someone running it in an hour. Power usage and batteries (stored work) are measured in kWh; solar panels, for example, are measured in kW (how much they *could* produce when the sun is blazing overhead). Typical energy draws: lights and computers in a house, 0.5 kW; 2-ton A/C unit, 1.7 kW; car charger, 3.5 kW; clothes drier on high heat, 5 kW. So, for example, charging the car for eight hours would consume 28 kWh (3.5 x 8). You can see how conservation matters a lot too.

26 Paul Martin. "What are the Energy Solutions?" LinkedIn Pulse, December 19, 2020, https://www.linkedin.com/pulse/what-energy-solutions-paul-martin/; Honest Government Ad (Satire) / Carbon Capture and Storage, https://www.youtube.com/watch?v=MSZgoFyuHC8.

27 James T. Tanner, "Black-capped and Carolina chickadees in the southern Appalachian Mountains," *The Auk* 69 (1952), 407-424; Brian Davidson, Michael J. Braun, and Brian Scott Davidson, "Cryptic genetic introgression into an Appalachian sky island population of black-capped chickadees," University of Maryland, 2011.

28 There are 7.5 gallons in one cubic foot of water. USGS data at Sault St. Marie, Michigan: https://waterdata.usgs.gov/nwis/dv?referred_module=sw&site_no=04127885.

29 Lake Huron and Lake Michigan are at the same elevation and thus could be considered the same lake. The relatively narrow strip of water at the Straits of Mackinac is the delineation between the two lakes.

30 Agricultural Marketing Resource Center (production map by locality): https://www.agmrc.org/commodities-products/fruits/cherries; National Cherry Festival: https://www.cherryfestival.org/p/about/383.

31 From an extensive phone interview with a third-generation family member of Island View Orchards.

32 Here is a good explanation of how heat is transferred efficiently to the oceans due to the inherent properties of water: https://www.climate.gov/news-features/understanding-climate/climate-change-ocean-heat-content.

33 National Oceanic and Atmospheric Administration, National Centers for Environmental Information, Climate Monitoring: https://www.ncdc.noaa.gov/sotc/snow/201701.

34 Here are two scientific articles on the subject of the California drought, which lasted for an excruciating five years (2012–2017): Noah S. Diffenbaugh, Daniel L. Swain, and Danielle Touma, "Anthropogenic warming has increased drought risk in California," *Proceedings of the National Academy of Sciences* 112 (2015), 3931-3936; D. Griffin and K. J. Anchukaitis, "How unusual is the 2012–2014 California drought?" *Geophys. Res. Lett.* 41 (2014), 9017–9023.

35 Jon Erdman, "Record flooding in April/May 2017 swamps parts of Missouri, Arkansas, Illinois," *The Weather Channel*, May 5, 2017, https://weather.com/storms/severe/news/flood-threat-forecast-south-mississippi-valley-april2017; USGS interactive map and monitoring station data: https://maps.waterdata.usgs.gov/mapper/index.html.

36 *Encyclopedia Britannica* has a good image of ice age extent: https://media1.britannica.com/eb-media/18/64918-004-69BCB037.jpg; "Shackelton, N.J. Oxygen isotopes, ice volume and sea level," *Quaternary Science Reviews* 6 (1987), 183-190. Here is a good video showing the advancement and retreat of ice

that was built from the type of science used in the above article: https://wgnhs.wisc.edu/wisconsin-geology/ice-age/.

37 Bradley Brooks (Indianapolis Museum of Art), "Athens of the Prairie: How Columbus, Indiana Got Great Architecture." Reason TV: https://www.youtube.com/watch?v=E8foQ_BC4sU.

38 There are a number of sources for calorie conversions, but here are my calculations.

Car:
30,000 = the number of equivalent food calories in one gallon of gasoline.
3,200 miles / 30 miles per gallon = 107 gallons.
107 gallons x 30,000 calories = **3,210,000** calories of energy burned in a continental car ride.

Bicycle:
I estimate that I pedaled on average 10 miles an hour.
3,200 miles / 10 miles an hour = 320 hours of pedaling.
650 calories per hour to propel a bicycle at 10 miles an hour x 320 hours = **208,000** calories.

See Endnote #14 for the efficiency of an EV compared to a gas car. Note: There is a basal metabolic rate (BMR) included in the 650 calories per hour that I would have burned, whether I was on the bike or sitting at home on a couch. I have disregarded BMR for simplicity. Subtracting basal metabolism would have lowered the bicycler's required calories to *only* propel the bicycle and would have made the bicycler look even more efficient than a vehicle.

39 There are about 87,000 commercial, private, and military flights a day in the United States, a third of which are commercial (i.e., public transport), https://sos.noaa.gov/datasets/air-traffic/. Here is the math: 10.7 million (population of Georgia) / 331.4 million (population of United States) = .032 x .17 (proportion of emissions from vehicles; see "Sources of US Greenhouse Gases"

on page 56) = 0.005; 1000 (daily flights from HJAIA) / 87,000 = 0.011 x 0.03 (proportion of emissions from jet airplanes) = 0.0003. But let's say that airplanes at HJAIA are using 1.5 times more fuel than the average jet (1.5 x 0.0003 = 0.00045); therefore, 0.005 / 0.00045 = 11 (rounded).

40 Raw data from the Mauna Loa Observatory: https://gml.noaa.gov/ccgg/trends/data.html; Tables and trends: https://www.esrl.noaa.gov/gmd/ccgg/trends/index.html.

41 https://climate.nasa.gov/evidence/.

42 Here is one source on seasonal atmospheric CO_2 fluctuations: https://scripps.ucsd.edu/programs/keelingcurve/2013/06/04/why-does-atmospheric-co2-peak-in-may/.

43 Jeremy D. Shakun et al., "Global warming preceded by increasing carbon dioxide concentrations during the last deglaciation," *Nature* 484, 49-54 (April 2012).

44 For a good description of ppm, see Skeptical Science, the non-profit science education organization: https://www.skepticalscience.com/CO2-trace-gas.htm.

45 For an insightful book on probability, see Leonard Mlodinow, *The Drunkard's Walk: How Randomness Rules Our Lives* (New York: Vintage Books, 2008).

46 Shantideva, *The Way of the Bodhisattva,* trans. by the Padmakara Translation Group, Shambhala Classics (Boston: Random House, Inc., 1997).

47 Global Issues is a media site edited by Anup Shah. Here is what he says: "My background is a degree in computer science—not exactly a degree in global issues! The point is that you don't need to have qualifications to be concerned and want to do something, although you do need the time to sift through a lot of information to understand what is happening!" http://www.globalissues.org/article/26/poverty-facts-and-stats.

48 Elise Gould and Hilary Wething, "U.S. Poverty Rates Higher, Safety Net Weaker Than in Peer Countries," Economic Policy

Institute, July 24, 2012, https://www.epi.org/publication/ib339-us-poverty-higher-safety-net-weaker/

49 Jared Diamond, *Collapse, How Societies Choose to Fail or Succeed* (New York: Penguin Books, 2005), 494-496.

50 The tax revenue (known as Transportation Special Purpose Local Option Sales Tax) is estimated to be $13 million over a five-year period for the City of College Park, http://www.fultoncountyga.gov/tsplost/.

51 M.S. Friedman et al. "Impact of changes in transportation and commuting behaviors during the 1996 Summer Olympic Games in Atlanta on air quality and childhood asthma," *Journal of the American Medical Association* 285(7), 897-905 (February 21, 2001).

52 Martin Moylan, Elizabeth Dunbar and Kirsti Marohn, "Xcel's new plan: Coal-free by 2030, nuclear until 2040," Minnesota Public Radio, May 20, 2019.

53 "U.S. Greenhouse Gas Inventory Report: 1990-2014," https://www.epa.gov/ghgemissions/us-greenhouse-gas-inventory-report-1990-2014.

54 Max Roser, "Why did renewables become so cheap so fast?," Our World in Data (Dec. 1, 2020).

55 Marty Ahrens, "Vehicle Fires," National Fire Protection Association (March 2020).

56 US Department of Transportation, National Highway Traffic Safety Administration, https://www.nhtsa.gov/press-releases/2020-fatality-data-show-increased-traffic-fatalities-during-pandemic.

57 Robert Channick, "Rivian plant gearing up for June launch, the return of automaking to Normal and the dawn of the electric vehicle age," *Chicago Tribune*, April 16, 2021.

58 "Fuel economy in aircraft," https://en.wikipedia.org/wiki/Fuel_economy_in_aircraft; http://www.boeing.com/resources/

boeingdotcom/company/about_bca/startup/pdf/historical/757_passenger.pdf.

59 Tina Hodges, "Transportation's role in responding to climate change," US Department of Transportation, Federal Transit Administration, January 2010, https://www.transit.dot.gov/sites/fta.dot.gov/files/docs/PublicTransportationsRoleInRespondingToClimateChange2010.pdf.

60 Georgia Power portfolio: https://www.georgiapower.com/company/about-us/facts-and-financials.html; MEAG portfolio: https://www.meagpower.org/power/generation/; New York power portfolio: https://www.eia.gov/state/?sid=NY; Leslie Kaufman, "How New York is building the renewable energy grid of the future," *Inside Climate News* (May 25, 2017), https://insideclimatenews.org/news/24052017/new-york-renewable-energy-electrical-grid-solar-wind-energy-coal-natural-gas.

61 Hodges, "Transportation's role in responding to climate change."

About the Author

John Duke is an entrepreneur who owns Hydroecology North America, Inc. He is the principal floodplain analyst and a subcontractor for small to midsized engineering and ecology firms in 12 states so far. If you have ever looked at a Federal Emergency Management Agency (FEMA) floodplain map, John may have helped create it. His work also includes modeling the restoration of wetlands for the US Army Corps of Engineers' wetland mitigation banks under the direction of the Clean Water Act.

He has no children of his own, but he married the widow, Marianne Lecesne, who did. Seven, in fact, including one adopted. He says that his wife is the hardest working person he has ever known. She holds down a career as a high-school calculus teacher and is the one who always mows the lawn—with an electric mower, of course. Except for two who come home from college at Christmas, John and Marianne are now empty nesters.

John often returns to Ely, Minnesota—especially in the wintertime—to walk, sleep, and fish on the ice along the Canadian border.

You can contact him at:
jd@hydroecology.org
linkedin.com/in/john-e-duke-60bab548

Made in the USA
Monee, IL
31 October 2021

80519439R20098